Economics of Aquaculture

Economics of Aquaculture

Curtis M. Jolly, PhD
Howard A. Clonts, PhD

CRC Press
Taylor & Francis Group
Boca Raton London New York

CRC Press is an imprint of the
Taylor & Francis Group, an **informa** business

First published 1993 by The Haworth Press

Published 2023 by CRC Press
Taylor & Francis Group
6000 Broken Sound Parkway NW, Suite 300
Boca Raton, FL 33487-2742

ISBN 13: 978-1-56022-020-6 (pbk)

Visit the Taylor & Francis Web site at
http://www.taylorandfrancis.com

and the CRC Press Web site at
http//www.crcpress.com

Library of Congress Cataloging-in-Publication Data

Jolly, Curtis M.
Economics of aquaculture / Curtis M. Jolly, Howard A. Clonts.
p. cm.
Includes bibliographical references and index.
ISBN 1-56022-020-1 (alk. paper).
1. Aquaculture industry. I. Clonts, Howard A. II. Title.
HD9450.5.J65 1922
338.3'713 – dc20

92-221
CIP

Dedicated to

PAULINE AND JEAN

Our wives who encouraged our efforts

ABOUT THE AUTHORS

Curtis M. Jolly, PhD, is Associate Professor in the Department of Agricultural Economics and Rural Sociology at Auburn University. He teaches courses in Aquacultural Economics, Project Planning and Sector Analysis, and Economic Development. He has over ten years of professional experience in teaching, research, and international development activities. Dr. Jolly, who has worked in many African and Caribbean countries, is presently engaged in research on sustainable agriculture, integrated aquaculture, and international trade. He is an associate of the International Center of Aquaculture and a member of the World Aquaculture Society and the American Agricultural Economics Association.

Howard A. Clonts, PhD, is Professor of Resource Economics in the Department of Agricultural Economics and Rural Sociology at Auburn University. He has served as a member of the Advisory Council for the International Center for Aquaculture at Auburn, and developed and taught the first courses in the economics of aquaculture at the University. Dr. Clonts' research interests include analysis of optimum water quantities for fish production, economics of fish pond effluent control for maintaining water quality, strategies for government action in aquacultural production in developing economies, and the economics of fish production. A member of the Southern Agricultural Economics Association and the Alabama Fisheries Association, he has over thirty years of experience in resource use and development.

CONTENTS

Foreword xi

Preface xiii

Acknowledgments xv

Chapter 1: Introduction 1

Growth of Aquaculture 3
Fish Consumption 9
Advantages of Aquaculture 12
Aquaculture in Economic Development 13

Chapter 2: Economics of Aquaculture 21

Economic Terminology 21
Functions of an Economic System 31

Chapter 3: Demand and Supply of Fish 35

Consumer Demand 35
Elasticity of Demand 42
Elasticity, Total and Marginal Revenue 54
Derived Demand 57
Producer Supply 58
Elasticity of Supply 63
Competitive Market Equilibrium 68

Chapter 4: Production 77

Purpose of Production 77
The Production Function 79
Relationship Among TPP, MPP, and APP 87

Optimizing the Use of a Single Resource 90
Empirical Production Functions 92

Chapter 5: Cost of Production **99**

Short-Run Production Costs 100
Simulated Cost Curves for Catfish 105
Long-Run Costs 109

Chapter 6: Factor-Factor and Product-Product Relationships **117**

How to Combine Inputs and Outputs 117
Profit Maximization 125
Imperfect Substitution 128
The Production of Two Outputs 133

Chapter 7: Farm Management **141**

Farm Planning 141
Developing a Farm Plan: Essential Steps 142
Economic Consideration of Management 145
Farm Operation 150
Farm Income and Budget Analysis 151
Income Statement 156
Depreciation 166
Partial Budgeting 170
The Enterprise Budget 173
Break-Even Analysis 176
The Balance Sheet 181

Chapter 8: The Time Value of Money **191**

Present Value Versus Future Value 191
Future Value of a Present Sum – Compounding 192
Present Value of a Future Payment – Discounting 196

Chapter 9: Capital Budgeting **203**

Net Present Value 205
Internal Rate of Return 208
Risk Management and Capital Budgeting 221
Risk Types 221

Risk Management 224
Quantitative Risk Measures 228
Utility Theory 232
Incorporating Risk into Capital Budgeting 233

Chapter 10: Market Structure and Theory of Price 237

Market Patterns 237
Monopolistic Competition 250

Chapter 11: Marketing 257

What is Marketing? 258
Marketing Functions 261
Marketing Channel 264
Integration 270
Catfish Marketing Channels in the Southern U.S. 272

Chapter 12: Government in Aquaculture 277

Typical Aquacultural Problems 278
Causes for Aquacultural Problems 279
Economics of Support Policies 287

Appendix 291

Author Citations Index 307

Subject-Author Index 309

Foreword

Growth and development of aquaculture as a technological innovation in food production has been rapid in recent years. This largely empirically-based food production system has responded positively to the application of the biological, physical, and social sciences. The result has been a growing scientific literature upon which to base management strategies, which in turn has improved production, marketing, and utilization of aquacultural products. Early in the maturation phase of technological innovation, questions of economic performance become paramount for further application and development of the technology.

Economic questions about aquaculture become exceedingly important as this technology plays on the world stage. Production from capture fisheries has stabilized, yet population growth is unabated in parts of the world. Fish, once the "poor man's food," is now priced beyond the reach of much of the world's poor. Protein malnutrition is a growing problem. Lack of disposable income limits options of many to improve the lives of their families. Accelerating natural resource degradation demands that sustainable food production practices be available to producers. Scarce land resources create pressure to seek more productive uses for agriculturally marginal lands. Aquaculture promises to provide relief to these concerns. However, the economic role of aquaculture must be understood as we push forward into the future. *Economics of Aquaculture* is an important tool in this effort. It is a reference for the professional and a text for the student. This is especially important since both professionals and students of aquaculture must learn to think in economic terms, and documentary resources and courses for this purpose are too few.

Beyond the basic production aspects of modern aquaculture is the ever growing need for integrated planning in food and resource management systems. This text presents the concepts necessary for a careful and systematic approach to planning. Government planners and policymakers should find this resource a valuable tool in guiding aquacultural development and understanding the place of aquaculture in an integrated food management system.

Economics of Aquaculture is applicable on an international scale. As Director for the International Center for Aquaculture and Aquatic Environments (ICA) at Auburn University, I have personally seen the benefits to colleagues worldwide which are traceable to the materials developed by Jolly and Clonts. Concepts presented in this text are an integral part of the training materials used at the ICA. They should be a part of similar materials whenever the science and art of aquaculture are presented.

Bryan L. Duncan, Director
International Center for Aquaculture
and Aquatic Environments
Auburn University
Auburn, Alabama

Preface

Aquaculture world is almost as old as human culture. Fish have been cultivated for centuries. Yet, it is still a relatively recent phenomenon in some parts of the developed world.

There is, now, more than ever, a growing awareness of the importance that food fish production may have with respect to human nutrition, employment, and even recreation in more developed societies. The need for nutritional improvement in less developed societies is well documented. Yet, there is often confusion in both developed and developing economies regarding the feasibility of producing fish as opposed to other "meat" products. The purpose of this book is to illustrate how aquaculture is growing in importance and how production feasibility may be determined.

The book should be of interest to students concerned with development of aquacultural activities, but it should also interest government planners who may be considering new ways to improve diets as well as foreign exchange. Thus, the future economist, fisheries biologist, and government planner should all find value in the materials presented.

The subject material flows easily from initial concepts on the role of aquaculture into production economics, management, marketing, and governmental policy. Basic economic principles for each are explained with ample illustrations. There is a mix of examples taken from the United States and other countries of the world. Thus, the student completing the study should be able to understand a wide variety of subjects and/or problems related to the economics of aquaculture.

Acknowledgments

The authors are deeply grateful for the invaluable assistance in the preparation of this book provided by several people. Among them are typists Beth Rush, Mary Anne Jones, Melanie Minor, and Sergio Duarte. Their dedication to an enormous task is appreciated.

The authors are indebted to their students who over the years have made useful suggestions and corrections for improving the manual. Particular mention should be made of two students, Mr. John Moehl and Ms. Joyce Newman, whose suggestions were of invaluable importance.

Professional guidance in technical matters was provided by several members of the faculty at Auburn University. The authors acknowledge Dr. E. W. Shell for his cooperation and moral support. Special thanks go to Drs. John Adrian and James Stallings for their review and comments. However, responsibility for the manuscript rests with the authors.

Chapter 1

Introduction

Fish accounts for nearly one quarter of the world's supply of animal protein, and in many developing countries it is the ideal and traditional supplement to a basic diet of starches (James, 1986). Fish has been described as the "meat" of the "Third World." Important contributions by the fisheries sector to both developing and developed economies are in the form of employment, income, and exchange earnings.

Growth in fish production increased from an annual catch of less than 20 million metric (20.32 million U.S.) tons in the late 1940s to exceed 65 million metric (71.6 million U.S.) tons by 1970. In the 1980s, about 80 to 90 million metric (81.28 to 91.44 million U.S.) tons of fish were produced (caught or cultured) worldwide annually (Rabanal, 1987; James, 1986). In 1987, world commercial fishery landings were about 93 million metric (94.48 million U.S.) tons. During the 1980s, landings increased consistently and by 1987 they were 29 percent above those of 1980. The former Soviet Block of nations combined and China have the largest landings, accounting for slightly over one-third of the world's commercial catch. The United States had the fourth largest catch in 1987 (USDA, 1989).

Underlying this expansion is the growth in the world economy, which was stimulated by two revolutionary technological developments. One was the introduction of synthetic fiber fishing nets and mechanical-gear hauling systems, which facilitated the use of large purse seines. The other was the introduction of freezing at sea, which allowed on-board processing and longer fishing trips (Bailey, 1987 and 1988). These innovations, along with the development of trawlnet fishing techniques, led to a dramatic increase in the size and the radius of fishing vessel operation. Figure 1-1 shows world

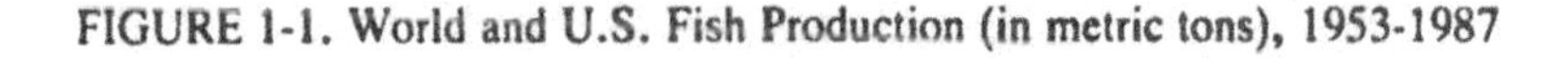

FIGURE 1-1. World and U.S. Fish Production (in metric tons), 1953-1987

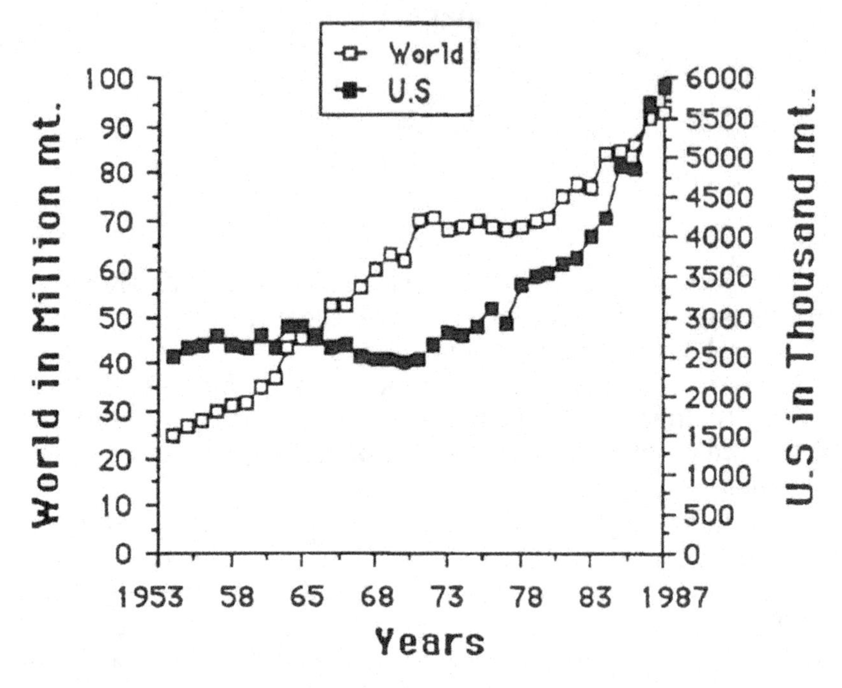

Source: The data were extracted from *Aquaculture Situation and Outlook Report*, USDA, September 1989. The world catch includes shellfish, but the data from U.S. include only finfish.

and U.S. fish production from 1953 to 1987. The solid points represent U.S. production in thousand metric (1.016 thousand U.S.) tons, while the open points represent world production in million metric (1.016 million U.S.) tons. Over the 35-year period, the overall annual growth rate for world production was 4 percent, while for 8 years, growth in the U.S. was 2 percent. However, if growth rate is calculated for the period 1975-1987, world production growth is less than that of the U.S.

The majority of fish consumed by humans comes from capture fisheries. The problem is that natural supply is reaching a plateau, while consumption is increasing. Though production increased for 10 consecutive years, there has been a recent decrease in the rate of growth. The 1987 landings were only 0.6 percent above that of 1986. As the rate of increase of marine fish harvests dropped to less than 1 percent per year in recent years, aquacultural yields which now account for approximately 10 percent of total fish production increased at a rate of more than 7 percent per year (Rabanal, 1987; Kent, 1986). Since 1975, world aquacultural output has increased from 4.5 to over 10 million tons in 1987, yet less than one-tenth of the potential sea and land surface area is utilized. Figure 1-2 shows world and U.S. aquacultural production (Mitchison, 1986).

> *Definition:* Aquaculture, which is commonly called fish farming, fish culture, or mariculture, is the rearing of fish, shellfish, and some aquatic plants, under controlled or semi-controlled conditions, for profit and/or human consumption. Freshwater aquaculture may include the farming of fin and shellfish, water chestnut, watercress, water-salad, bull frogs, prawns, crayfish, and other shellfish. Another form of aquaculture, mariculture, is the farming of sea fin and shellfish, such as flatfish, seabasses, mollusks, and shrimp, and the husbandry of marine organisms.

Techniques used in aquaculture may increase the production of fish and other aquatic foodstuffs far above the level that would be produced naturally. Aquaculture may be conducted in fresh water, brackish water, or seawater and also in flooded fields and rice paddies. The fish may be confined in earth ponds, concrete ponds, cages suspended in the open seas or lakes, or in impounded coastal waters. Confinement means the fish are protected from many natural predators but are more susceptible to disease and other problems.

GROWTH OF AQUACULTURE

Fish culture began in 1000 B.C. in China and probably even earlier in Egypt and Assyria. The first known Chinese monograph on the subject of fish farming dates back to the fifth century B.C. Fish breeding was first found attractive by the Chinese. The selective breeding of ornamental goldfish was later introduced to Japan

FIGURE 1-2. World and U.S. Aquacultural Production, 1953-1987

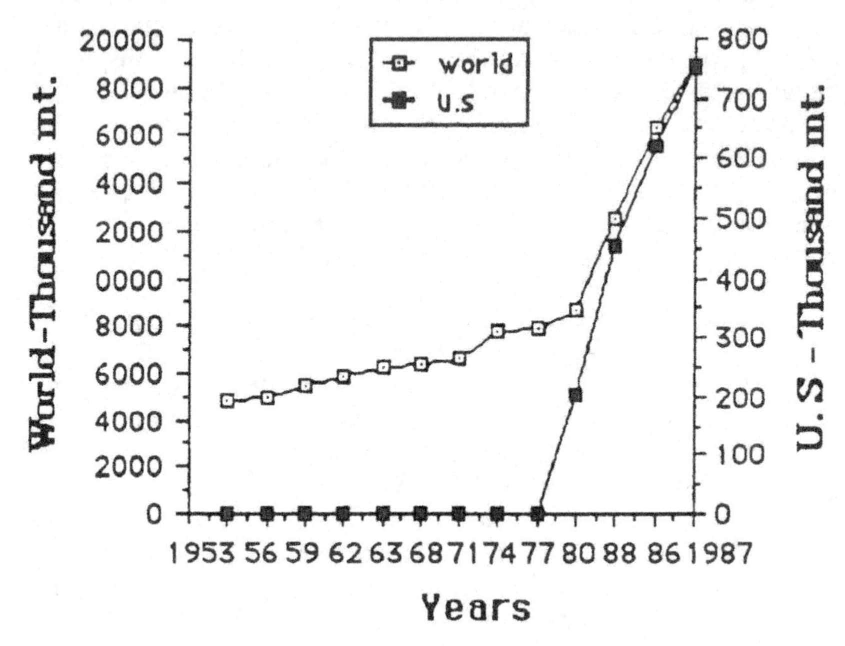

Source: The data were taken from *Aquaculture Situation and Outlook Report*, USDA, September 1989; and FAO Fisheries Aquaculture Production (1984-1987) Circular 815 No. 1, 1989.

where ornamental carp breeding was perfected. The ancient Romans, who kept fish for food and entertainment, were the first known marine aquaculturists. They constructed ponds that were supplied with fresh seawater from the ocean. By about 1400 A.D.,, brackish water fish farms had been established in Java, and the common carp, *Cyprinus carpio*, had reached Europe from Asia transforming fish culture there (*Encyclopedia Britannica*, 1988). About the same time, four other species of carp, including the grass carp, *Ctenopharyngodon idellus*, and the black carp, *Mylopharyngodon piceus*, joined the common carp in mixed culture ponds in China. The collection and compilation of statistics on world aquacultural production is limited. It is, however, estimated that world production will approach 16.5 million metric (16.76 million U.S.) tons by 1992. This growth is a direct result of increased financial investments in intensive aquacultural farming. China is the leading producer of finfish and other aquacultural products. In 1987 China was responsible for over 50 percent of all finfish cultured. Figures 1-3a and 1-3b show 1987 world aquaculture production in metric tons. East Asia is the overall largest producer of aquacultural products.

United States

Commercial warm water fish farming in the United States was begun in the 1920s and early 1930s by a few persons who raised minnows to supply the growing demand for baitfish for sport fishing. In the 1950s, producers mainly in the southern states, began raising a number of species such as catfish, buffalo, bass, and crappie as food fish. Today, the major component of U.S. aquaculture comes from the channel catfish industry. Figure 1-4 shows that the catfish industry is the largest contributor to the aquacultural industry.

Fish production in the U.S. more than tripled between 1975 and 1983, from 130 million pounds to over 400 million pounds. Catfish have shown the most rapid expansion in production. The principal edible aquaculture products (by weight) in the U.S. are catfish,

FIGURE 1-3a. World Aquaculture Production 1987 (in metric tons)

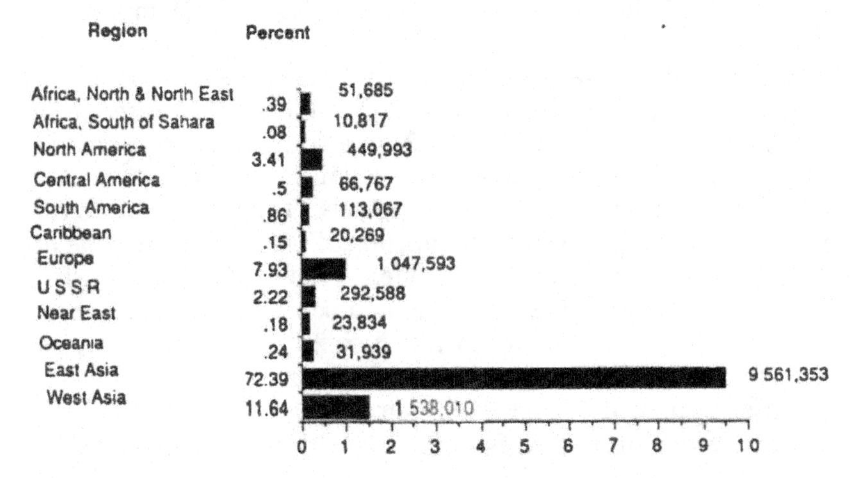

FIGURE 1-3b. World Finfish Aquaculture Production 1987 (in metric tons)

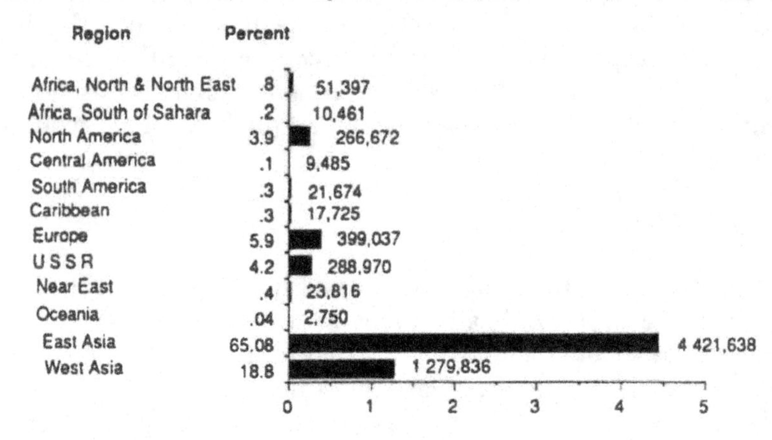

Source: Production data were taken from Nash, C. E. and C. B. Kensler (1990), and "A Global Overview of Aquaculture Production in 1987," *World Aquaculture*, 21(2): 104-112.

crawfish, salmon, and trout. These four species account for more than 90 percent of the aquacultural products moving through commercial channels. Catfish culture in the United States accounted for 73,840 acres of water in 1982 and reached 1 million acres by 1990. Commercial and noncommercial production of catfish reached more than 388 million pounds in 1988, about 4 percent above 1987. The value of this production was estimated at $321 million for 1988.

United States crawfish production in 1987 was about 105 million pounds, with a value estimated at $53 million dollars. Production levels for salmon and trout for 1987 were 80 and 59 million pounds, respectively. Total aquacultural production from all species in 1987 was approximately 746.9 million pounds. Though catfish constitutes 50 percent of all private production of fish, it is responsible for only 44 percent of the total fish value, as shown in Figure 1-4. Bait fish account for only 3 percent of total production, but for 9 percent of total value. Another species which contributes more in percentage of total value than its production is oysters.

Mississippi remains the largest producer of fish in the U.S., where the dominant species is catfish. Other major producers of catfish are Alabama, Arkansas, Louisiana, and Texas. Figures 1-5 and 1-6 show the number and average size of commercial catfish operations by state. Alabama has the largest number of farmers engaged in catfish production, but most of the farmers produce on relatively small acreages of less than 50 acres. Mississippi is the leading state in terms of average size of operations. Arkansas is second in terms of average size of operation. In all other states, farmers produce on farms of less than 50 acres. In Alabama, 12,400 acres of water were devoted to catfish production in 1986, most of which was located in the west central area of the state. In 1989, there were about 14,275 acres—an increase of about 14.3 percent.

Africa

In Africa, various species of Oreochromis, loosely referred to as tilapias, have been cultured since 1943. The tilapias are known for rapid breeding in ponds, a wide feeding spectrum, and ease of culture. Therefore, they serve as a good food fish, especially under limited resource conditions. In 1987, Africa south of the Sahara, which includes 35 reporting countries, produced 10.8 million metric (11.9 million U.S.) tons of aquacultural products whereas the North and Northeast, which include Algeria, Egypt, Morocco, and Tunisia, produced 51.7 million metric (52.5 million U.S.) tons.

Asia

Asian countries are the oldest producers of aquacultural products, and today are the leading producers of many species. East Asian countries dominate with 72 percent of all production. According to Nash and Kensler (1990), this domination is due to their massive production of finfish and seaweeds. West Asian countries produce 12 percent of the total world production. In 1987, 3 million metric (3.04 million U.S.) tons of seaweeds were produced in East Asia alone, while the production in West Asia was 60,000 metric (60.960 million U.S.) tons. These countries also produce the largest quantities of aquacultural shellfish.

Europe

European countries produce a substantial amount of aquacultural products. Production in 1987 was estimated at 1 million metric (1.016 million U.S.) tons, comprising 3,285 metric (3337.6 million U.S.) tons of crustaceans and 655,595 metric (655,595 million U.S.) tons of mollusk. European countries contribute 8 percent of the total world output of aquacultural products.

Latin America and the Caribbean

Fish production through aquaculture in Latin America and the Caribbean has been relatively recent. The quantities produced are only a small portion (1.5 percent) of the total world production. The industry is, however, growing in importance. Many Latin Ameri-

FIGURE 1-4. U.S. Aquaculture Production Distribution by Species in Percent, 1987

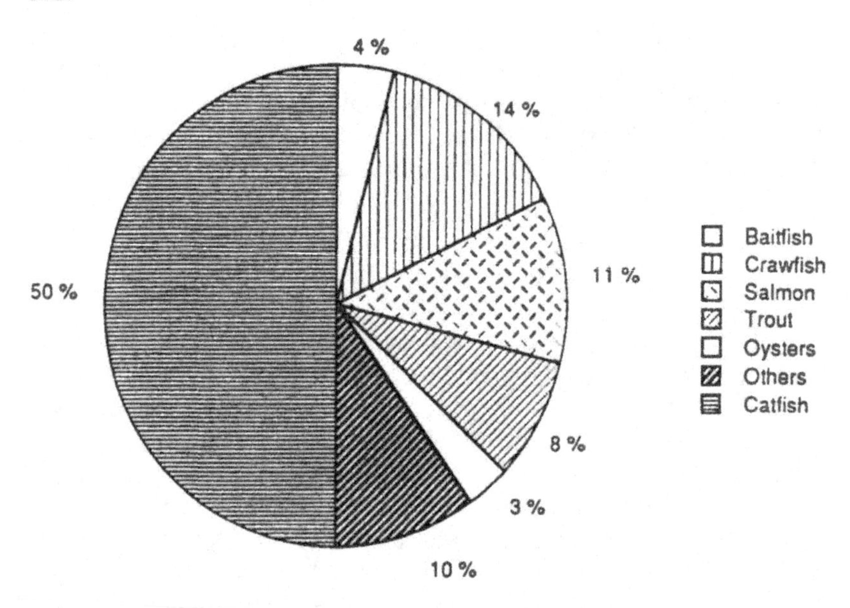

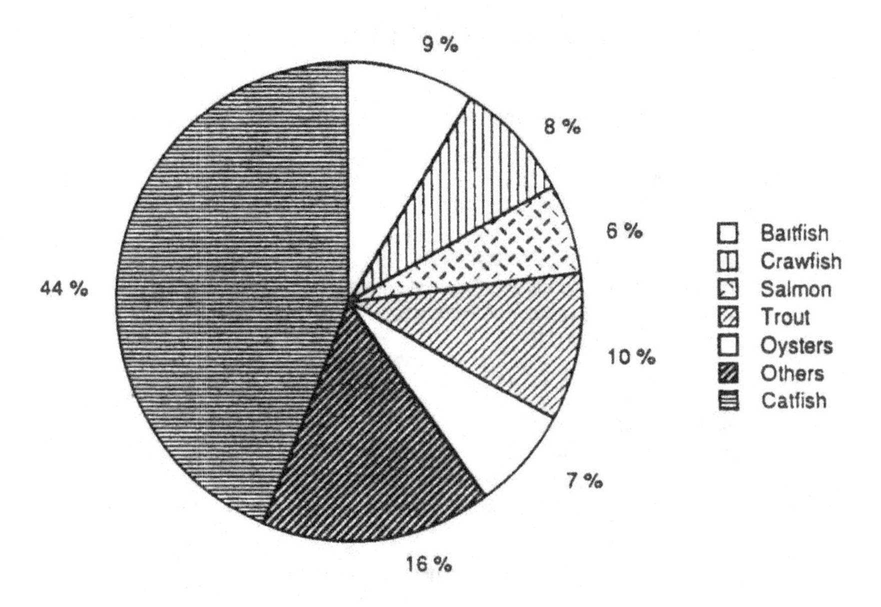

B - Value of Fish Produced

can countries concentrate on prawn and shrimp production for export.

FISH CONSUMPTION

Developing countries account for about half of the world fish production, but are responsible for a little less than half of the world's fish consumption (Kent, 1986). It must be borne in mind that population in all developed countries together amounts to only a third of the population of the less developed countries (James, 1986). Thus, the consumption differences are much more significant when per capita measures are used.

Fish, being rich in protein, is well suited to human dietary requirements. It compares favorably with eggs, milk, and meat in nutritional value of protein and amino acid composition. Lack of sufficient protein of high nutritional value is a common nutritional

FIGURE 1-5. Number of Catfish Operations in 1989

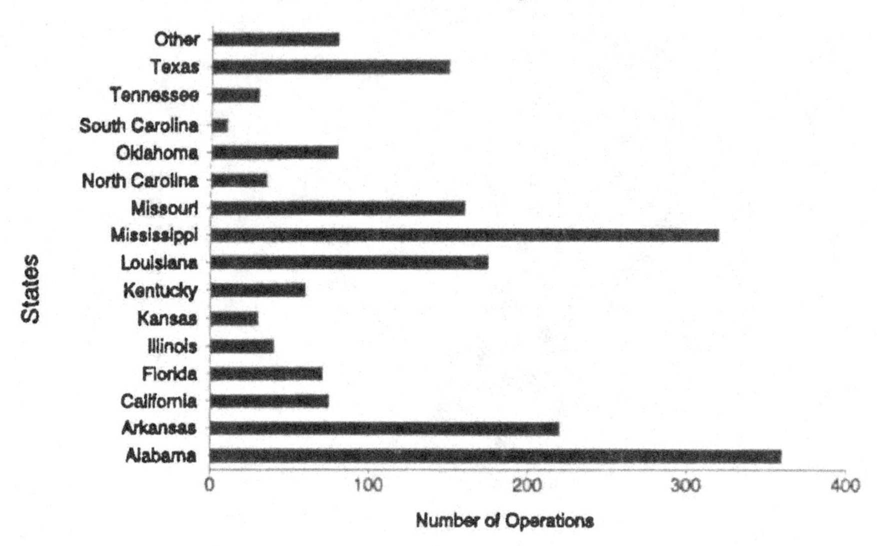

Source: Data were taken from USDA Agriculture Situation and Outlook *National Agriculture Statistics Service, USDA, 1989.*

deficiency in many tropical countries. An increased supply of processed fish may help alleviate this situation.

Less developed countries are highly dependent on fish as a source of protein, as shown in Figure 1-7. However, as the figure shows, as income increases, the relative preference for fish declines and that for red meat increases somewhat. According to the FAO, 40 percent of the developed world population relies on fish as a source of protein, whereas 45 percent of the developing world depends on fish as a source of protein (FAO, 1980). Thus, consumption in the third world has accounted for much of the world fish demand. Over the past 30 years, per capita world fish consumption remained fairly constant in developed countries while that of developing countries increased 2 percent. It is also evident that the households of lower socio-economic strata spend more of their income on fish than on meat. In the poorest societies of the world, preserved fish furnishes a significant amount of the total dietary animal protein. For remote inland communities, aquaculture provides a nutritious and year-

FIGURE 1-6. Average Size of Catfish Operation

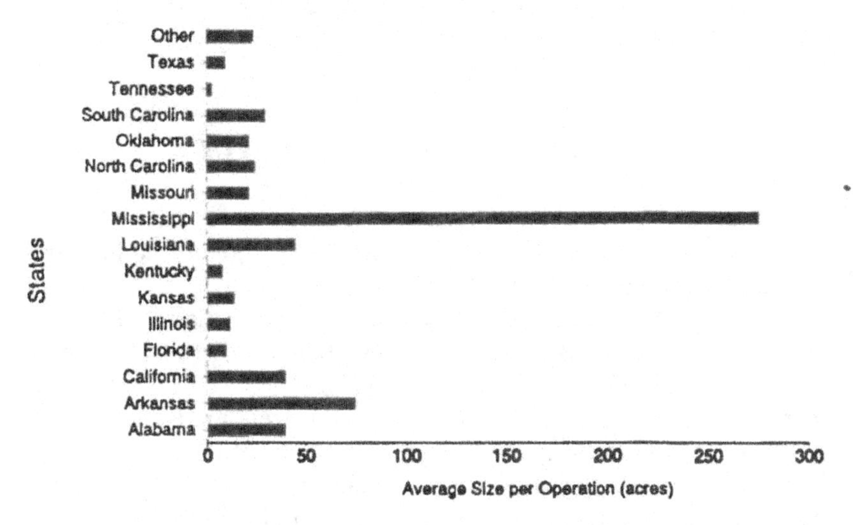

Source: Data were taken from National Agriculture Statistics Service, USDA, 1989.

long source of good quality protein. Previously, only coastal communities had this rich protein source.

Fish Consumption and Health Effects

Many studies have confirmed the health benefits of increased consumption of fish and fish products. Research thus far suggests that populations with the highest consumption of fatty fish appear to have the lowest incidence of cardiovascular diseases. Epidemiological studies have confirmed an inverse relationship between fish consumption and cardiovascular disease among Greenland Eskimos, whose intakes average nearly 400 grams (14 ounces) of fish and arctic mammals per day (Bang et al., 1980). Low incidence of cardiovascular disease has also been observed among Japanese whose per capita fish consumption averages 113 grams (4 ounces) per day (FAO, 1980). A study published by Kromhout et al. (1985) suggests that intakes of fish as low as 30 grams (1 ounce) per day or as little as one to two fish entrees per week may be sufficient to reduce the mortality risk of coronary heart disease. Fish consumption has also been linked to reduced hypertension, reduced blood clotting tendencies, and more favorable plasma lipid and lipo-protein levels (Pavelec, 1989).

ADVANTAGES OF AQUACULTURE

Fish culture has existed on a commercial basis for centuries, but has rarely achieved the dominance enjoyed by agricultural methods of producing animal protein. By virtue of the relative importance of fish in their diet and an extensive history of fish culture, the Japanese and Southeast Asian nations rank high among those in which fish culture is a significant industry (Liao, 1988). In recent years, there has been a pronounced increase in interest in fish culture in North America. In part, this interest reflects a concurrent increase in public attention regarding aquatic resources, particularly marine resources (Pavelec, 1989). It also reflects publicity given to allegedly superior food conversion efficiency of aquatic species and their productivity per unit of growing area. Proponents of fish culture often point out the superior biological efficiency of fish over land animals. The buoyancy of water helps fish to expend less energy than land animals for support and movement. Fishes are also cold-blooded and expend less energy on body temperature maintenance than warm-blooded animals. Low food conversion of fish feed to fish protein frequently is cited as an advantage of fish culture systems. The protein efficiency ratio or the conversion of dietary protein to flesh is about the same for fish and chickens, but is substantially better than for swine or cattle. Lovell (1989) estimated that the feed-to-food ratio is 1.9:1 for catfish; that is, 1.9 pounds of feed produces 1 pound of catfish product. The ratio of hogs, cattle, and poultry is 4:1, 8:1, and 2:1 respectively.

Fish may also grow in ponds without commercial or supplemental feed. This is common in milkfish culture throughout Southeast Asia. Usually organic and inorganic fertilizers are added to the pond to increase the production of zooplankton, a basis for the food chain. In many other developing countries, compost from animal and farm wastes are used for pond fertilization. The primary cost is labor for transporting the compost.

Aquaculture also contributes to economy in several ways. Pond culture of fish:

1. provides a food source for a growing population;
2. provides a protein source to humans (which is essential for body growth, repair, and reproduction);
3. provides a limited amount of employment and income for people;
4. utilizes some of the nation's waste products;
5. contributes to foreign exchange; and
6. makes use of land and other resources unsuitable for agriculture and other industries.

AQUACULTURE IN ECONOMIC DEVELOPMENT

Approximately 85 percent of the 750 million people in less developed nations are considered poor based on the arbitrary criterion of an annual per capita income equivalent of $50 or less. Of the population in less developed countries considered to be either absolutely or relatively poor, more than 80 percent live in rural areas. Agriculture is the principal occupation of four-fifths of the rural poor. The need for special intervention to increase rural food production and incomes applies also to the provision of social and other services, such as health and education. Poverty is reflected in poor nutrition, inadequate shelter, and low health standards. These affect not only the quality of life, but also the productivity of rural people.

The common response to the problem of food and nutrition has been to grow more food. More recent analysis, taking account of low demand or purchasing power among the poor and malnourished, has also emphasized the need to increase their incomes.

Historically, the past agricultural production in developing countries concentrated on large-scale production for export markets. While there was growth in the economy and the people employed in large-scale production received a wage, malnutrition persisted because the wage earners could not afford to buy sufficient food to maintain themselves and their families. The attention today is to produce more food for the domestic market at a low price so that

FIGURE 1-7. Relative Expenditure of Food Share Budget (in percent) for Fish and Meat by Ranked Income Groups in Less Developed Countries

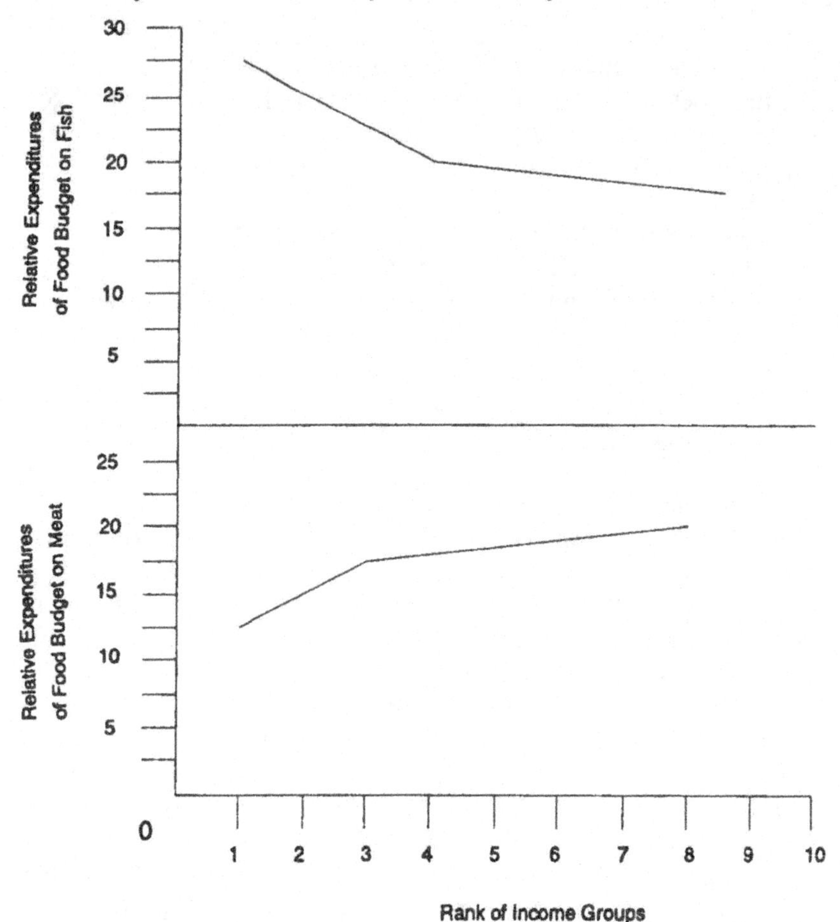

Source: Data were taken from FAO consumption statistics, 1974-1985. The income groups were made of 15 less developed countries from Africa, Asia, the Caribbean, and Latin America.

most of the population can afford to purchase the increased production.

In most developing countries, the rural poor have developed skills in producing rice, cassava, edoes, sorghum, millet, potatoes, or other carbohydrate-rich foods. However, the ability to produce protein-rich foods is limited. Though aquaculture is no panacea for malnutrition, it has great potential. World production of protein from aquaculture is less than desirable. In order to solve the problem of malnutrition in rural areas, knowledge in fish production is vital to rural people. Poor people cannot depend solely on marine fisheries to solve the problems of malnutrition, low income, and unemployment because:

1. marine fish catch is low and decreasing at an increasing rate in many developing nations;
2. poor countries cannot afford the equipment to exploit the fisheries resources along their coasts;
3. the marine resources of many developing countries are overexploited through indiscriminate fishing by local and international fishermen with superior equipment;
4. even the fish caught by developing nations cannot reach the poor people since there are inadequate storage, processing, and handling facilities; and
5. because of the lack of processing and storage facilities, only the people close to the coastline areas benefit from marine fisheries.

Aquaculture can be one of the solutions in providing the basic needs of the rural poor. A source of protein can be provided to the rural poor at an affordable price. Also, it may generate limited employment for the poor people in rural areas, and provide a major boost to local economic development. These provisions depend, of course, on the species cultured and the intensity of operations.

The pond culture of fish may aid the efficient utilization of one of the most limiting resources in developing countries, the land, since pond culture may be implemented on lands marginal for agricultural activities. Fish may be produced on salt lands, coastal flats, mangrove swamps, and other low productivity lands (FAO, 1954). Inland fisheries activities may be recommended to less developed countries (LDCs) because of the direct economic advantages as well as their importance in other areas such as social stability.

World Fishery Trade

World fish trade is very important to developing and developed countries. Developed economies are the main exporters and importers, but the value of exports contributes significantly to the gross domestic product (GDP) of many developing countries. Total value of world fishery product imports was estimated at $30.5 billion in 1987. Japan is the world's largest importer at $8.3 billion. The United States is second with $5.7 billion. Canada is the world's largest fishery products exporter at $2.1 billion in 1987. The United States, Denmark, Korea, and Norway are the next largest exporters with values of $1.8, $1.8, $1.5, and $1.5 billion, respectively (USDA, 1989).

The growing demand for fisheries products has led many countries to engage in aquaculture to earn foreign exchange. High priced species such as shrimp, prawns, salmon, trout, and miscellaneous shellfish products are the primary export products. The quantity of aquacultural products is still relatively small as a percentage of total fishery product export, but it is still gaining in importance. International trade in aquacultural products is expected to accelerate because the demand for fisheries products is increasing faster than the catch from ocean fisheries.

Limitations of Aquaculture

The importance of aquaculture to rural development is situational. The feasibility of aquaculture development efforts depends on local marketing conditions, available natural resources, and social attitudes. Market prices for fish may be insufficient to stimulate farmers to invest in aquaculture for a number of social, financial, and technical reasons. Thus, unless the fish can be marketed at a reasonable price, farmers are unlikely to invest in aquaculture. Natural resources such as land and water must be available in quantity and quality to have a successful aquacultural project. Also, the people's attitudes must favor aquacultural development before any aquacultural project is planned. There are several conditions which facilitate the successful implementation of an aquacultural project (Ben-Yami, 1980). These include:

1. a favorable market for the intended species;
2. a positive attitude regarding aquaculture and the willingness of people to accept changes brought about by a new industry;
3. a political will to enable equitable access of the prospective fish farmers to natural resources (land, water);
4. ready availability of seed, feed, equipment, materials, disease prevention and treatment, extension services, credit, and financial markets favorable to fish farming; and
5. convincing indicators of economic benefits to the prospective fish farmers and their families.

REFERENCES AND RECOMMENDED READINGS

Bailey, C. 1988. "The Political Economy of Fisheries Development in the Third World." *Agriculture and Human Values*. 3(4):35-48.

Bailey, C. 1987. "The Social Consequences of Tropical Shrimp Mariculture Development." *Ocean and Shoreline Management*. II:31-44.

Bailey, C. and S. Jentoft. 1990. "Hard Choices in Fisheries Development." *Marine Policy*. 14(4):333-44.

Bailey, C., D. Cycon, and M. Morris. 1986. "Fisheries Development in the Third World: The Role of International Agencies." *World Development*. 14(10/11):1269-75.

Bang Ho, J. Dyerberg, and H.M. Sinclair. 1980. "The Composition of the Eskimo Food in North Western Greenland." *American Journal of Clinical Nutrition*. 33:2657-61.

Ben-Yami, M. 1980. "Aquaculture: The Importance of Knowing its Limitations." *The FAO Review, Food and Agriculture Organization of the United Nations*. Rome, Italy, 19,4: 15-9.

Campbell-Asselbergs, E. 1986. "A Nutritional Approach to Fisheries Projects, in Food and Nutrition." *Food and Agriculture Organization of the United Nations*. Rome, Italy, 12(2):36-8.

Campbell-Asselbergs, E. 1986. "Research for Fisheries and the Alleviation of Malnutrition, in Food and Nutrition." *Food and Agriculture Organization of the United Nations*. Rome, Italy, 12(2):38-43.

Courtenay, W.R., Jr., and J.R. Stauffer, Jr. 1984. *Distribution Biology, and Management of Exotic Fishes*. Baltimore: The Johns Hopkins University Press, pp. 1-40.

Coutts, D.C. 1986. "Fish and Food Aid, in Food and Nutrition." *Food and Agriculture Organization of the United Nations*. Rome, Italy, 12(2):28.

Food and Agriculture Organization of the United Nations. 1989. *Aquaculture Production (1984-1987)*. Fisheries Circular 815 No. 7.

Food and Agriculture Organization of the United Nations. 1980. *Food Balance Sheets and Per Capita Food Supplies*. Rome, Italy: FAO; 1980.

Food and Agriculture Organization of the United Nations. 1954. *Fish Farming and Inland Fishery Management in Rural Economy*. Rome, Italy, pg. 64.

Food and Agriculture Organization of the United Nations. 1981. *The Prevention of Losses in Cured Fish*. Rome, Italy, pp. 1-5.

Food and Agriculture Organization of the United Nations. 1984. *A Study of Methodologies for Forecasting Aquaculture Development*. Rome, Italy, Technical Report 248, pp. 1-29.

George, T.T. 1975. "Introduction and Transplantation of Cultivable Species into Africa, Supplement I to the Report of the Symposium on Aquaculture in Africa," *Food and Agriculture Organization of the United Nations*. Rome, Italy, pp. 407-9.

Grover, J.H., D.R. Street, and P.D. Starr. 1980. "Review of Aquaculture Development Activities in Central and West Africa." *Research and Development Series*, No. 28; International Center for Aquaculture, Agricultural Experiment Station, Auburn University, Alabama. pg. 31.

Hjul, P. 1988. "The Future of Aquaculture." Proceedings of Aquaculture International Congress and Exposition, Vancouver Trade and Convention Center, Vancouver, British Columbia, Canada; September 6-9. pp. 3-6.

James, D.G. 1986. "The Prospects for Fish for the Malnourished, in Food and Nutrition." *Food and Agriculture Organization of the United Nations*. Rome, Italy, 12(2):20-7.

Jolly, C.M. 1989. *Economics of Aquaculture Production and Farm Management*. Compiled and Formatted by T.J. Pfeiffer and I. Echeverry, International Center for Aquaculture, Fisheries and Allied Aquaculture, Auburn University, Alabama. 60 pages.

Kent, G. 1986. "Impacts of Fisheries Policy, in Food and Nutrition." *Food and Agriculture Organization of the United Nations*. Rome, Italy, 12(2):32-5.

Kromhout, D., E.B. Bosschieter, and C.L Coulander. 1985. "The Inverse Relationship Between Fish Consumption and 20-Year Mortality From Coronary Heart Disease." *New England Journal of Medicine*. 312:1205-9.

Lartey, B.L. and S. Dzidzienyo. 1986. "Fisherwomen in Africa, in Food and Nutrition." *Food and Agriculture Organization of the United Nations*. Rome, Italy, 12(2):19.

Liao, I. 1988. "East Meets West: An Eastern Perspective of Aquaculture." *Journal of the World Aquaculture Society*. 19(2):62-73.

Lovell, T. 1989. *Nutrition and Feeding of Fish*. New York: Van Nostrand Reinhold. pg. 260.

Maar, A., M.A.E. Mortimer, and I. Van Der Lingen. 1986. *Fish Culture in Central East Africa*, Food and Agriculture Organization of the United Nations, Rome, Italy, pg. 1.

Meltzoff, S.K. and E. LiPuma. 1985/1986. "The Social Economy of Coastal Resources: Shrimp Mariculture in Ecuador, in Culture and Agriculture." *Bulletin of the Anthropological Study Group on Aquarian Systems*. pp. 1-10.

Menasveta, P. 1987. "Present Status and Potential Uses of Biotechnology for Increasing the Coastal Aquaculture Production in Thailand." *Journal of Aquaculture in the Tropics*, 2:107-26.

Mitchison, A. 1986. "Aquaculture: Still Promising." in Food and Nutrition, *Food and Agriculture Organization of the United Nations*. Rome, Italy, 12(2):30-2.

Nash, C.E. and C.B. Kensler. 1990. "A Global Overview of Aquaculture Production in 1987," *World Aquaculture*. 21(2):104-12.

Parker, N.C. 1984. "Technological Innovations in Aquaculture." *Fisheries*. 9(4):13-16.

Pavelec, C.L. 1989. "Fish Consumption and Omega-3 Fatty Acid Intake in Central Alabama Men With and Without a History of Heart Disease." Master of Science Thesis, Auburn University, Auburn, Alabama.

Rabanal, H.R. 1987. "Managing the Development of Aquaculture Fisheries." *Food and Agriculture Organization of the United Nations*. Rome, Italy.

Rupp, E.M, F.L. Miller, and C.F. Baer. 1980. "Some Results of Recent Surveys of Fish and Shellfish Consumption by Age and Region of U.S. Residents." *Health Physics*. 39:165-75.

Sandifer, P.A. 1988. "Aquaculture in the West, A Perspective." *Journal of the World Aquaculture Society*. 19(2):73-84.

Shang, Y.C. 1986. "Status, Potential and Constraints to Development of Coastal Aquaculture in Asia." *Infofish Marketing Digest*. (5)10-3.

Stickney, R.R. 1988. "Commercial Fishing and Net-Pen Salmon Aquaculture: Turning Conceptual Antagonism Toward a Common Purpose." *Fisheries*. 13(4):9-13.

The New Encyclopaedia Britannica, 1988, Vol. 1, 15th Edition S.V. "Aquaculture."

USDA, 1989. *Aquaculture Situation and Outlook Report*. Economic Research Service, AQUA-3, p. 34.

Chapter 2

Economics of Aquaculture

Economics is concerned primarily with the study of problems associated with the production and distribution of economic goods and services. This means that economists are interested in *what* is produced, as well as how and for whom. Thus, economics is the study of the proper method of allocating scarce physical and human resources among competing ends. Economics deals with choices between alternatives. Problems of choice arise only when resources are scarce and alternative uses can be made of them.

ECONOMIC TERMINOLOGY

Scarcity

Scarcity is generally loosely defined as a limited supply of a good or service. In a stricter sense it implies the limited supply of a *desirable* good or service. Air to breathe is desirable, but it is generally considered unlimited and therefore is not scarce. Clean air, however, is scarce. Garbage seems to be unlimited, but is not desirable. Therefore, it is not scarce. Clean drinking water, on the other hand, may be both desirable and limited. Hence it may be scarce.

People's wants are diverse and insatiable, but the resources for producing the things they want—land, labor, raw materials, factories, buildings, machinery—are themselves limited in supply. There are insufficient resources in the world to produce the amount of goods and services necessary to satisfy everyone's wants fully. Consequently all things that are desirable at all times are said to be "scarce." The real meaning is that goods and services are scarce relative to the demand for them.

Choice

Scarcity is created by the relative desire for goods and services, brought about by unsatisfied wants. Since all wants cannot be satisfied, choices must be made. If every human action was perfectly rational, everyone would naturally satisfy his or her more pressing wants first, choosing the things most desired, and going without those considered less desirable. By and large, this may be so, though many people have given way to impulse and made a purchase they afterwards regretted.

The question of choice is fundamental to economics. How many times is the complaint heard that, "I do not have time to go to the farm to cut the weeds invading the fish ponds?" Generally, however, this is merely an excuse for not doing something that is disliked. What is really meant is that something really liked is done instead, for example, sleeping late. Time is scarce and one must choose between competing ways of spending it.

The greatest amount of choice occurs, however, in the expenditure of one's income. First, one has to decide how much to invest in aquaculture and how much to save, or how much to spend on rent, food, clothing, and holidays. People often say they cannot afford to buy something when they really mean they prefer to spend their money on something else.

Scales of Preference

One of the fundamental assumptions of economics is that man is a rational being who makes rational decisions. Thus, if a person is forced to choose between one thing or another, it is assumed that the alternative chosen will always be one that will yield the greater satisfaction. This implies, too, that each individual has a preference scale, a list of all unsatisfied wants arranged in order of preference.

A fish farmer is faced with the problem of choosing between purchasing feed and/or chemicals. The farmer is restrained by a budget. A government officer may have to decide between investing in large-scale marine fisheries or large-scale aquaculture: Like the fish farmer, the government officer also is constrained by a budget, and may choose to produce some fish through investing in small-scale marine fisheries and some through large scale aquacultural production.

More often, the decision is not between two alternatives, but among an infinite number of possibilities. Figure 2-1 demonstrates a useful way of viewing economic choice at the national level. We imagine a simple economy producing only two types of products, aquacultural and (marine) fisheries. We see two areas, one attainable (dotted area) and one unattainable area. The dotted area is bounded by a production possibility curve, or production frontier. If all resources were devoted to small-scale fisheries, OA amount of fish would be produced. If all resources were devoted to aquaculture, OB amount of aquacultural products would be produced. If resources are divided between the two methods of producing fish, various combinations of aquacultural and fisheries products may be produced. Point E_1, for example, represents a combination of OA_1 fisheries products and OB_1 of aquacultural products. Point E_2 shows a combination of OA_2 fisheries products and OB_2 of aquacultural products. If we increase the production of fish through investment in large scale aquaculture we have to reduce investment in small scale fisheries.

The slope of the production possibility curve is concave downward. The curve illustrates the cost of increasing the amount of one commodity in terms of sacrificing consumption of the other goods. Or stated alternatively, it is the value of one item in terms of the other. In other words, it measures the *opportunity cost*, or the value

FIGURE 2-1. Production Transformation Curve

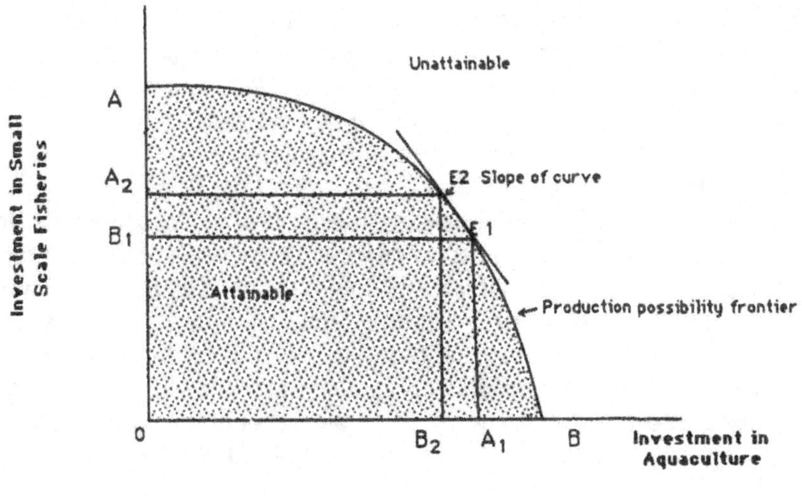

of the opportunity foregone. Opportunity cost then is a result of resource scarcity.

Definitions in Economics

> Economics is the study of the allocation of scarce resources to meet relatively unlimited human wants and needs. Those resources which are scarce are termed *economic goods*.

If the aspect of scarcity is not present in a resource then it does not represent an economic problem area. Resources which can be obtained with ease and at no cost are called *free goods*.

Economics is not concerned with specifying which ends are "good" and which are "bad." Those subjects are in the domain of ethics, culture, and religion. The economist may say only that, given the means available, one end is in conflict with another, or, that given the end, this choice of means or resources allows maximum attainment of the end. *Positive Economics* is that body of economics that is concerned with "what is." This body of economics sees its task as providing a system of generalizations that may be used to make correct predictions about the consequences. Positive economics is based on facts. *Normative economics* deals with "what to do" or what "ought" to be done, and is related to morals, ethics, or social preferences which are not motivated by price conditions.

It is customary to divide economics into two parts: "micro" economics and "macro" economics. *"Macro" economics* deals with the behavior of economic aggregates such as gross national product, the level of employment, and the various sectors of the economy (for example, the fishing, or aquacultural sector), while *"micro" economics* deals with the economic behavior of individual units such as consumers, farmers, resource owners.

The prefix "macro" comes from the Greek word meaning "large." Therefore, macroeconomics means "economics in large." This is in contrast to microeconomics which can be interpreted as "economics in small." When one analyzes the output and the price of that output for a single firm (or industry), one is engaged in microeconomic analysis. On the other hand, macroeconomics involves analysis of the output level for the whole economy. For example, the average wage of all workers employed throughout the economy is the stuff of macroeconomics. There is primary concern with the study of aggregate economic activities without identifying separate firms or individuals. Typically, macroeconomics describes, analyzes, and forecasts economic activity, then evaluates resource allocation to meet developmental priorities.

Macroeconomic Tools

In order to measure whether a country is progressing toward satisfying more of its population's diverse needs, we need to know the volume of goods and services that can be produced through the economic activities. One such aggregate measurement of economic progress is called the Gross National Product or GNP.

Definitions of GNP and some other key terms used in macroeconomic analysis are in order at this point.

> *Gross National Product* (GNP) is defined as the total market value of all final goods and services produced in the economy in one year. All goods produced in a particular year may not be sold; some may be added to inventories. Nevertheless, any increase in inventories must be included in determining GNP, since GNP measures all current production regardless of whether it is sold.

The GNP represents the money value at market price of all final goods and services produced by a nation during a year. The GNP is calculated by multiplying quantities by *market prices*. Where a country's economy is based on barter or subsistence, its GNP will likely be underestimated. This means that such a country will appear poorer than it really is. GNP comprises only *final* goods and services. All consumer and capital goods for investment are considered final. In contrast, raw materials that are intermediate are excluded from the GNP calculation since they will appear as part of the value of final production.

Problems appear when we try to measure economic activity in the real world. There are problems of fitting concepts to reality and of getting accurate estimates. Another problem is distinguishing between nominal and real values.

Nominal measurements are made in terms of the price prevailing at the time measurements are made. Nominal amounts are not very useful for economic analysis because they can increase either when people buy more physical goods and services—more cars, steaks, fish—or simply when prices rise. Nominal measurements are adjusted for changes in the price level to estimate real values. A real magnitude is the value expressed in the price of an arbitrary chosen "base year." If the base year is 1972, real 1981 consumption "in 1972 prices" represents the amount that actual 1981 purchases would have cost if one were able to buy each item at its 1972 price. For instance, if all prices doubled between 1972 and 1981, fish purchases of $10,000 in 1981 would have cost only $5,000 at 1972 nominal prices. But, real consumption in 1981 measured in 1972 prices is $10,000. The ratio of nominal to real GNP is given a special name called the ***Implicit GNP Deflator***. The deflator tells us the ratio of prices in any single year.

The GNP can be compared with a cake. This cake may be either eaten or saved. In practice, part of the cake is eaten, part of it is saved. The savings are used for investment, which can be in the form of capital investment or stock. So, for a country with no foreign trade, GNP is the sum of consumption plus investment.

Further definitions of the sub-parts of GNP are in order. The term "investment" in economics means capital formation or creation. Capital is a produced good used in the production of a final good. For example, an aerator is a product made for sale to those who need it. Yet, the purpose of an aerator is to facilitate the production of fish. A tractor may be used in the production of food fish; the aerator and tractor are capital, the food fish are not. Anytime there is an increase in the ability to produce more tractors or aerators, there is investment. Thus, investment means deferring current consumption in order to use resources to produce more for consumption later. Economists say that new investment is *capital formation*.

The use of the term gross national product means that in calculating output of each producing unit, no allowances were made for the replacement of capital goods during the year because of obsolescence or depreciation (wearing out). If amounts of money necessary to replace capital are subtracted when measuring gross national output, the result is *Net National Product* (NNP). **NNP is GNP minus capital consumption or depreciation.** Note: There is no specific mention of capital above that amount needed for replacement, but it is included in NNP. NNP is also sometimes called national income at market prices, or the total sales value of output after deducting capital consumption.

National income at factor cost is the amount which is attributable to the factors of production. It is the value used if one is interested in tracing the changing importance of labor income relative to property income. Definitions for factors of production are provided later.

The amount of income that households receive is called *Personal Income*. Households do not receive all the income generated by the factors of production. Some is siphoned off before it reaches a nation's households. The government usually takes a portion through the corporate (business) income tax, and the board of directors of a business firm may vote to hold the remainder. Personal income includes household income received from every source, including transfer payments, before payment of personal income tax.

Disposable Income is simply personal income after payment of personal income taxes. It is how much money people finally have left to spend or to save. If disposable income is expected to increase, people involved in business, including fish farmers and other individuals in the fisheries sector, will expect the value of product sales to increase.

The average level of prices for individual goods and services changes frequently. Changes in relative prices are caused by changes in supply and demand. Changes in supply and demand are caused by changes in the society, such as tastes or conditions of production. The variability in savings over time at the average price level creates difficulties in measuring total income. National income is the product of quantities of goods and services times prices. Thus, national income increases during a particular year may be the result of either an increase in price or quantity produced.

Consumer Price Index (CPI) is a measure of the average level of prices for commodities purchased by a moderate-income urban family. The CPI is derived from weighing current prices by the average quantities of goods and services purchased in the base year and dividing this sum by the cost of those quantities in the base year. In the United States, approximately 400 of the most important commodities are sampled monthly by the Bureau of Labor Statistics, which computes and publishes the CPI. A sample method for calculating the CPI is seen in the Appendix.

The *Wholesale Price Index (WPI)* is the index of prices charged for goods sold in primary wholesale markets. Wholesale markets refer to basic goods produced in manufacturing, agriculture, forestry, fisheries, mining, and electric and gas utilities. As with the CPI, the WPI is in percentage terms relative to some base time period.

Economic Systems

Generalized names have been given to types of economic systems based on the method of resource allocation. The type of political system and type of economic system are not necessarily related. Nonetheless, every economic system must resolve various decisions:

1. What goods and services will be produced?
2. Who will be engaged in production?
3. What resources must be used in producing the goods and services?
4. How will the production be distributed?
5. For whom will the goods and services be produced?

The economic system will be judged by whether it produces the highest "standard of living" which its available resources and techniques will permit. That means the economic system ensures that the benefits of producing certain types of goods and services are available to maximize welfare for all of society's members. This is translated as equitable distribution of income which in turn connotes that great disparities between society's richest and poorest are unacceptable. A model of an economic system is shown in Figure 2-2. The model shows the flows of goods, services, and transactions from one unit in the system, say household, to another (firm).

Market Economy

A competitive market is one in which prices are determined solely by the free play of supply and demand. An economy characterized entirely by such markets would be a pure market economy, sometimes called a "competitive economy." It has the following primary characteristics:

1. A competitive economy is one in which the consumer is king, i.e., consumer sovereignty prevails. This means that consumers "vote" by offering relatively more dollars for products that are in greater demand and relatively fewer dollars for products in lesser demand.
2. It allocates resources efficiently. The struggle for survival forces producers to be efficient by offering the goods at prevailing market prices. Any producers who are unable to offer their product at the "going" market price within the industry will eventually be driven out of business.
3. It assures substantial economic freedom. A pure market economy requires a relatively high degree of individual freedom of enterprise and economic choice. Unlike a planned economy, there is no central authority who decides what, how, and for whom economic resources should be used. In a competitive economy, these decisions are made individually by producers as they seek to earn profits by allocating resources according to the way consumers freely register preferences through the price system.

Disadvantages of the Market Economy

1. Competition strains the abilities of businesspersons. It encourages a breakdown of competition. Consequently, it encourages them to avoid competition either by conspiring among themselves to monopolize markets and fix prices, or by engaging in various unhealthy forms of unfair competition such as "cutthroat" (below-cost) pricing, false advertising, bribery, and similar practices.
2. It fails to accord with modern technology. The model of a pure market economy makes the unrealistic assumption that industries are composed of a large number of small firms as envisioned by Adam Smith, the so-called Father of Economics. Smith saw production being carried out by a myriad of independent firms, all acting as if guided by an invisible hand, working to achieve maximum profit with maximum efficiency in resource use. The modern purely competitive model is based on Smith's concepts.

Planned Economy

In centrally planned economies, a central planning authority decides what will be produced and ultimately available for consumption. Between the two extremes of the completely free market-economy or the rigid, centrally planned economy, there are many intermediate economic systems.

Mixed Economic Systems

Most economic systems are neither purely market or planned economies. In general they are a mix of both systems. A mixed economic system exhibits the characteristics of free enterprise economy in much of its activity, but some decisions are made by the central government. In most mixed systems some decisions about housing, infrastructure, and public services are planned and operated by the central authorities. Governments may go even further by purchasing

FIGURE 2-2. Model of an Economic System

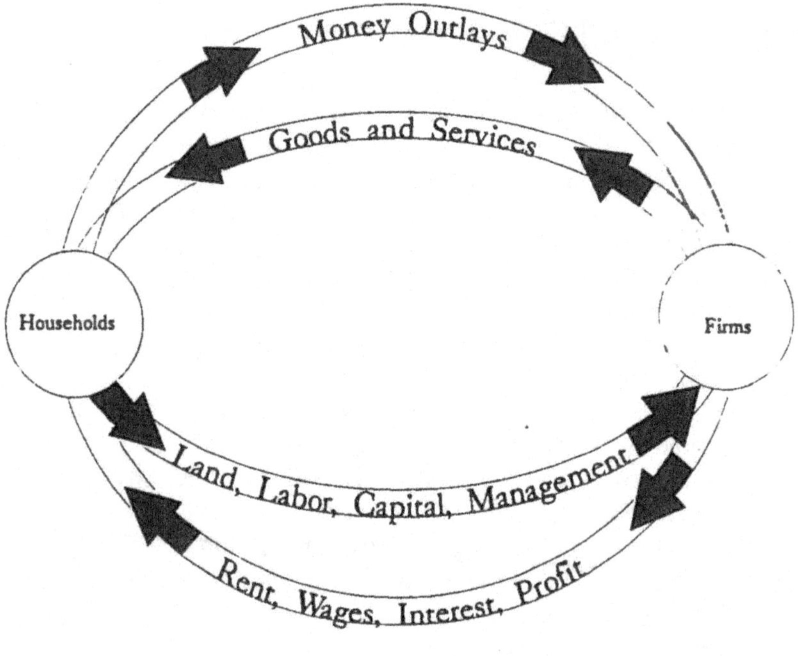

Money Outlays
Goods and Services
Households
Firms
Land, Labor, Capital, Management
Rent, Wages, Interest, Profit

more than 50 percent of the capital share in essential businesses such as mining and manufacturing. This is done to ensure that the government maintains some degree of control in those industries.

Governments with mixed systems usually promote equitable distribution of income for the society and thus provide a legal framework consistent with economic freedom. This means they attempt to prevent abuses of economic power. These governments insist on the development of policies that will ensure that there is a minimum supply of public goods which are needed for economic stability and growth.

FUNCTIONS OF AN ECONOMIC SYSTEM

The functions of an economic system have already been discussed in general, but now there is a need to be more specific.

1. *What to Produce.* This is the problem of determining the wants and needs in the community: which are most important, and in what degree they should be satisfied.
   a. The economy must have a method of establishing values for different goods and services that is acceptable to the group and reflects the relative desires of the group for the goods and services that the economy can produce.
   b. The value of an item to society is measured by its price in a private enterprise economy. The value is determined by buyers as they spend their income.
   c. The more urgent the need for a good, the more willing consumers are to back up desire with money and higher prices. The greater the supply of a particular good, the lower its price.
   d. This situation tells us what to produce based on a market, not what *ought* to be produced, based on ethics. A consumer with a larger income will exert more influence on the price structure than a consumer with lower income, but larger masses of people with lower incomes likewise will exert market influence. Market volume is as important as price.
2. *Organization of Production.* This involves (a) drawing resources from industries producing goods that consumers value

less and channeling them into industries producing favored goods, and (b) efficient use of resources by individual firms.

a. The price system in a free enterprise economy organizes production.

b. The term *efficiency* is a relation between input and output. Economic efficiency is measured in monetary terms of higher profit or lower cost.

c. Inputs (resources) will be used in such a manner to achieve maximum economic efficiency.

3. *Output Distribution.* Distribution of product is simultaneous with decisions of what to produce and organization of production. Individuals with higher incomes obtain larger shares of product.

   a. Income for an individual depends on three things:

   – the quantity of different resources that can be put into production;

   – the price of resources;

   – the value of an individual's labor relative to prices of goods purchased.

   b. Income distribution depends on the distribution of resource ownership in the economy and whether individuals put resources to work producing what consumers want most–where higher prices are.

   c. Income differences that arise from improper channeling of resources into production tend to be self-correcting.

   d. Income differences arising from differences in resource ownership *will not* be self-correcting.

   e. Society may impose forced corrections in resource allocation or resource ownership through tax, subsidy, or institutional changes, e.g., land reform.

4. *Short-run Rationing.* An economic system must make some provision for rationing commodities over the time period during which supplies cannot be changed. This is called short-run or market period rationing. Example: grain supply, an annual crop, is fixed in supply. Two things are needed in allocating output. First, supply must be spread among consumers. Then the supply must be stretched over a time period until the next harvest. In private enterprise, *price* allocates goods among consumers. Price will rise to the point where all consumers

together will clear the market. *Price* also tends to allocate over time. It is low initially and high at the end of the time period as scarcity increases due to product consumption.

5. *Economic Maintenance and Growth.* Every economy must maintain and expand its productive capacity.
   a. Maintenance – keeping the productive power operating through provisions for depreciation.
   b. Expansion – continuous increase in kind and quantity of nation's resources, together with continuous technological improvement.

For countries that have some degree of national planning, there are at least three possible phases of economic planning:

1. the macro phase;
2. the sectoral phase; and
3. the project phase.

The *macro phase* sets a desirable target for the entire economy and indicates how much total investments each sector should receive. In the *sectoral phase*, an examination is made as to whether the investments allocated to the sector are *feasible*, in light of each sector's absorptive capacity. The sector plan determines which geographical areas and population segments deserve priority. Finally, the tentative project identification forms a link with the *project phase* planning cycle, where individual investment projects are analyzed. Project planning addresses whether the project is feasible from technical, economic, financial, and institutional points of view. After feasibility is confirmed, the investments may be undertaken.

The market economy is a coordination of many interlocking markets. Macro-level economists study aggregate market behavior. Micro-level analyses are made of individual markets for goods and factors of production. Markets are linked to sectors of the economy. By studying markets in one sector, one can grasp the essence of the behavioral pattern in the other sectoral markets. In this book, markets in the agriculture sector, particularly the aquacultural sub-sector, are studied.

Aquaculture Economics

For purposes of this book, the aquaculture sub-sector is differentiated from the agriculture sub-sector. Thus, aquaculture economics is defined in the context of the aquatic environment in which the discipline exists.

> Aquaculture economics deals with rearing of desirable aquatic organisms under controlled or semi-controlled conditions for economic or social benefits. Of special concern are the allocation and utilization of scarce resources (land, labor, capital, and management) in the production of aquatic organisms under managed conditions to satisfy some human want.

REFERENCES AND RECOMMENDED READINGS

Albrecht, W.P., Jr. 1983. *Economics*. Fourth Ed. Englewood Cliffs, New Jersey: Prentice Hall.

Arya, J.C. and R.W. Lardner. 1985. *Mathematical Analysis for Business and Economics*. Englewood Cliffs, New Jersey: Prentice Hall.

Browning, E.E. and J.M. Browning. 1989. *Microeconomic Theory and Applications*. Third Ed. Glenview, IL: Scott, Foresman, and Company.

Clark, C.T. and L.L. Schkade. 1974. *Statistical Analysis for Administrative Decisions*. Cincinnati: South-Western Publishing Co.

Darby, M.R. 1976. *Macro-economics: The Theory of Income, Employment and the Price Level*. New York: McGraw-Hill Book Co.

Ferguson, C.E. and S.E. Maurice. 1974. *Economic Analysis*. Illinois: Richard D. Irvin, Inc.

Freund, J.E. and B.M. Perles. 1974. *Business Statistics*. Englewood Cliffs, New Jersey: Prentice-Hall, Inc.

Havrilesky, T.M. 1980. *Introduction to Modern Macro-economics*. Illinois: Harlan Davidson, Inc.

Hirshleifer, J. 1980. *Price Theory and Applications*, 2nd Ed. Englewood Cliffs, New Jersey: Prentice-Hall, Inc.

Reynolds, L.G. 1985. *Macroeconomics: Analysis and Policy*. Illinois: Richard D. Irvin, Inc.

Simon, J.L., 1975. *Applied Managerial Economics*. Englewood Cliffs, New Jersey: Prentice-Hall, Inc.

Williams, H.R. 1978. *Macroeconomics: Problems, Concepts, and Self-Tests*. Third Ed. New York: W.W. Norton and Co.

Wonnacott, P., 1984. *Macroeconomics*, Third Ed. Illinois: Homewood.

Chapter 3

Demand and Supply of Fish

The majority of production decisions that a fish farmer makes are influenced by the forces of *supply* and *demand*. For example, the decision of what to produce is determined by the questions on whether the product is saleable as well as the individual farmer's preference. The determinants of a farm's profitability are the demand for its product and any production costs which affect supply. However, no matter how efficient the production process, and how cleverly a farmer manages a fish pond, or even what product is preferred, the business cannot be operated profitably unless a demand for the product (fish) exists.

Consumer demand is defined as the various quantities of a particular commodity a consumer is willing and able to buy as the price of the commodity varies, when all other factors affecting demand are held constant. Notice here, the phrase, "*a consumer is willing and able to buy*." To want fish is not the same as to demand it. *Effective demand* is the term used to distinguish wants and needs from demand.

The amount of fish taken from the market under all the imperfections that exist in any specific situation is *existing demand*. The demand relationship may be viewed as specifying the maximum quantity of fish an individual or group desires and is able to purchase at a given price; or the maximum price per unit that an individual or group is willing and able to pay for a given quantity of fish.

CONSUMER DEMAND

Demand Schedule

Market or consumer demand is a reflection of the expressed willingness to purchase goods and services according to prevailing price conditions. Thus, each individual has trade-offs regarding market price and quantity purchased. Hence, a schedule of prices and quantities may be determined for that individual. For normal goods, when price is high, fewer units of the good are purchased. When price is lowered, a greater quantity is taken.

When the entire market is considered, the expressed willingness to buy of all individuals is summed for each price level. That is, the quantity individuals A, B, or C will buy at various prices is added to determine the market quantity.

Table 3-1 represents a demand schedule for the quantity of fish taken at different prices. A look at the schedule shows that the lower the price of fish, the greater the quantity that is demanded. The data from Table 3-1 are used to plot the downward sloping, individual demand curve on Figure 3-1. The demand curve is generally negatively sloped from left to right.

Demand Curve

The demand curve is a graphical picture of the demand schedule. Thus, the demand curve for a particular commodity may be defined as a locus of points, each of which shows the maximum quantity of the commodity that will be purchased in a given market, at a given time, at a particular price. The curve represents an attempt to relate a rate of quantity flows to a price at an instant of time.

The *Law of Demand* states that more of a commodity is taken

TABLE 3-1. Hypothetical Demand Schedule for Fish

Quantity of Fish Purchased in lbs.	Price $
5	5.00
10	4.00
15	3.50
20	3.00
25	2.50
30	2.00
35	1.50
40	1.00

FIGURE 3-1. Hypothetical Demand Curve for Fish

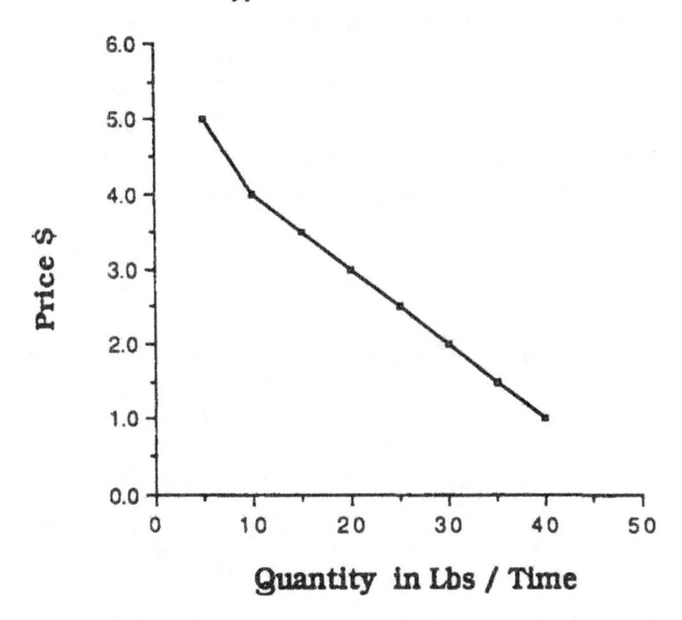

when its relative price falls, all other things remaining constant. In other words, price and quantity demanded are inversely related. This assumes that in reality consumers have perfect knowledge of the market for the commodity. However, a distinction exists between *existing demand* and *theoretical demand*. *Existing demand* is the quantity of a product taken from the market under all the imperfections that exist in any specific situation. *Theoretical demand* is the amount of the product that would be purchased under idealized circumstances if numerous buyers and sellers had perfect knowledge of prices and goods availability, and adequate supplies of competing, supplementary, and complementary goods.

Demand and Quantity Demanded

A distinction must be made between "demand" in the schedule sense and "demand" in the sense of quantity demanded. The term demand is used when reference is made to the non-price forces that shape an individual's demand. The term *quantity demanded* is used when reference is made to a particular quantity at a particular price.

That is, we are moving along the demand curve. Quantity demanded is a *static* concept in that quantity response to price is considered, and all other factors (incomes, personal tastes and preferences, and the prices of competing products) are assumed constant.

Change in demand over time is a *dynamic* concept. Changes in demand are usually associated with changes in income, population, or other variables influencing demand and which occur with the passage of time. A change in any of these variables may result in a change in buying patterns even though price remains constant. Thus, reference is made to a *change in demand*.

Factors Affecting the Demand for Fish and Fish Products

Food fish is considered exceptionally valuable from a nutritional standpoint, primarily because it contains a high percentage of readily digestible animal proteins. The demand for fish, however, is determined by many other factors in addition to nutritional considerations. Fish might only be an intermediate product in the industry and its demand will be influenced by conditions in the end-product markets. The factors influencing the demand for fish may be grouped under four headings:

1. population size and its distribution by age and geographic area;
2. consumer income and distribution;
3. consumer taste and preferences; and
4. prices and availability of substitutes for fish.

Population Size and Distribution

Population trends have an impact, not only on total demand, but also on per capita demand, because of differences in consumption patterns between regions, sexes, age groups, family units of different size, religion, and social traditions. The population distribution may give an indication of the number of fish lovers, or the number of people in a category, who consume fish and the amount of fish that is consumed by each group. In the United States it is expected that demand for fish would be highest in areas highly populated with peoples of Asian decent. The per capita consumption of fish for Asian peoples is higher than that for other ethnic groups (Pavelec, 1989).

Religious and cultural practices in a given society are powerful determinants of demand. Bell (1968) showed the effects of religious beliefs on fish consumption. For example, the Catholic Church once required members to abstain from eating meat during the period called Lent. Because fish is considered to be a good substitute for meat, this practice has the effect of increasing the demand for fish during such periods. It is possible to show quantitatively the substantial magnitude of shift in demand for fish during Lent. There is clear evidence that during this period, individuals substituted fish for meat thereby shifting their demand curves for various species of fish to the right.

The demand curve D_0 D_0 in Figure 3-2 shows a shift to the right to D_1 D_1. The shift of the demand curve illustrates the effects of population characteristics on demand. The demand for fish was shifted because of the number of Catholics found in the New England States, and the availability of substitute commodities. In 1966, American Catholics ended mandatory meatless Fridays. This

FIGURE 3-2. Illustration of a Change in Demand

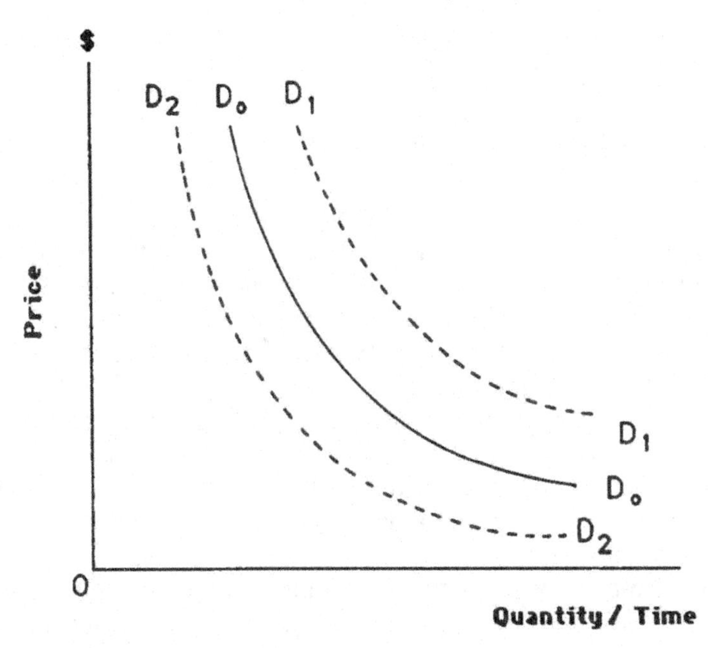

change had the effect of reducing the demand for various species as some individuals substituted meat for fish. When the Catholic prohibition against eating meat on Fridays ended in December 1966, the consumption of cod fell by 29.0 thousand pounds. The consumption of large haddock fell by as much as 7.1 thousand pounds after December 1966 when the rules were relaxed. This reduction in fish consumption and the accompanying shift in demand can be represented by D_2 D_2 in Figure 3-2.

In spite of the change in restrictions on meat consumption during Lent, some Catholics still continue eating fish on Fridays. Dellenbarger et al. (1988), in a study conducted in Louisiana, found that Catholic households had greater expenditures for catfish than those of other religious preferences. In that study they also found that occupation had no significance in explaining expenditures for catfish; however, race did, as blacks had higher expenditures for catfish.

Rupp et al. (1980) indicated that age and location of residence were major factors in the quantity of fish consumed. The frequency of fish consumption increased with age. Also, people living in the coastal regions were found to eat more shellfish and marine finfish than those residing in central regions of the United States. Residents living in the central regions consumed more freshwater fish than the national average.

Consumer Income and Distribution

Income and demand are positively related. That is, an increase in income typically shifts the demand curve to the right. Yet, for a few commodities the reverse is true. If the food is a normal good then an increase in income will bring about a shift in the demand curve to the right. *Question:* What does this say for foods that an individual may consider inferior?

In relatively high-income countries, per capita income has comparatively less impact on food consumption levels than in low-income countries. The physical quantity of all food consumed per person, for example, is relatively fixed in high-income countries. Demand studies indicate that in many developed countries the quantity of fish consumed relative to red meat has tended to decline.

Changes in demand may also occur as income is redistributed

from the rich to the poor. It is possible to increase the demand for trout or prawns by transferring income to families near or below the poverty line without changing the total or average level of income. In the United States, food stamps make it possible for lower income families to purchase trout, prawns, or more expensive species. Kinnucan and Wineholt (1989) found that consumer income had an unclear effect on catfish demand. The negative income effect they mentioned was due to an image problem acknowledged by the industry. Catfish was often viewed as a low income food commodity, or an inferior good. The study by Dellenbarger et al. showed more clearly that the income-expenditure elasticity of catfish indicated that fish consumption decreased with increased levels of income.

Prices and Availability of Substitutes

The demand for a commodity is a function not only of its own price, but of prices of many other commodities and services. A change in the price of one commodity may bring about a shift in the demand for other commodities. If, for instance, the price of mullet falls relatively lower than the price of catfish, then the demand curve for catfish will be shifted to the left.

For complementary products, the change in the price of related commodities and the change in demand are usually inversely related. Results of various demand studies indicate substitution among red meat, poultry, and fish, although the influences of substitutes are rather weak.

Consumer Tastes and Preferences

Consumer preferences are formed over a long period of time and reflect local availability and regularity of supply delivery, and the form in which fish are offered for sale. Shifts in preferences and demand may occur when a new condition of supply availability persists over a long period, or when the consumer becomes acquainted with a new product. Since food habits do not change easily, new products can seldom be introduced without substantial advertising and other promotional efforts.

The swing in developed countries to improved diets with lower cholesterol content has been one of the driving forces in demand shifts of seafood and aquacultural products. More healthy diets involve some replacement of red meats, which are higher in saturated fat content, with poultry and fish, which are lower in saturated fats.

ELASTICITY OF DEMAND

Price Elasticity of Demand

The elasticity of demand is the degree of responsiveness of quantity demanded to changes in price. At each point on a market demand curve the price elasticity of demand is different. The mathematical definition of price elasticity of demand is:

$$E_p = \frac{\frac{\Delta Q}{Q}}{\frac{\Delta P}{P}} = \frac{P}{\Delta P} \times \frac{\Delta Q}{Q}$$

where:

Q = quantity of the good
P = price of the good
Δ = a very small change.

An alternative equation for defining price elasticity is arc formula:

$$E_p = \frac{\frac{Q_0 - Q_1}{Q_0 + Q_1}}{\frac{P_0 - P_1}{P_0 + P_1}} = \frac{Q_0 - Q_1}{Q_0 + Q_1} \times \frac{P_0 + P_1}{P_0 - P_1}$$

> The price elasticity-of-demand coefficient (E_p) for any commodity may be interpreted as the percentage change in quantity due to a percentage change in the price of that commodity, other factors held constant. Elasticity does not determine the **shape of demand curve, but there are some shape relationships.**

If demand is perfectly elastic, the demand curve will be a straight line parallel to the base line (Figure 3-3). If demand is perfectly inelastic the demand curve will be a straight line at right angles to the base and parallel to OY (Figure 3-4). Between these two are an infinite number of possibilities. A relatively elastic demand may be represented by a gently sloping demand curve (Figure 3-5), and a relatively inelastic demand may be represented by a steeply sloping demand curve (Figure 3-6). However, one must remember that slope is not the same as elasticity.

FIGURE 3-3. Perfectly Elastic Demand Curve

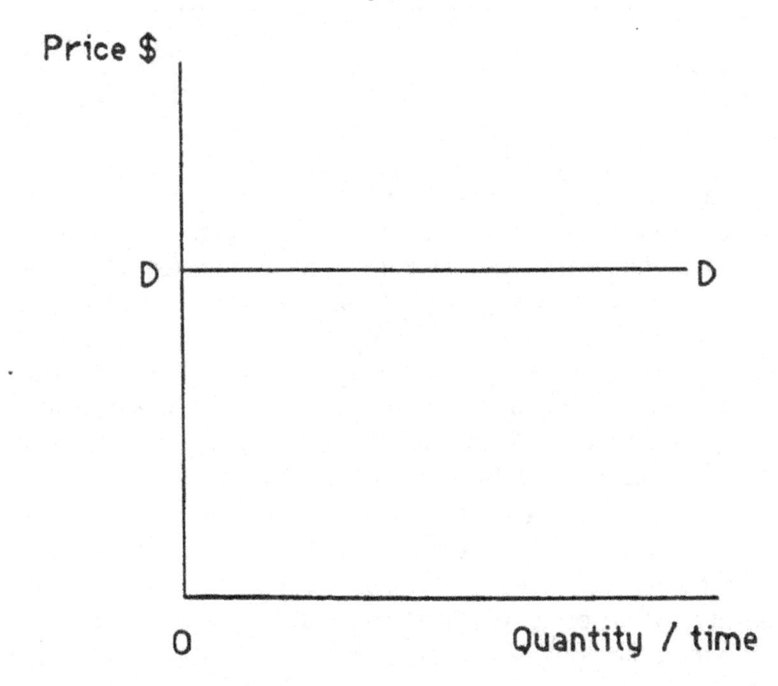

FIGURE 3-4. Perfectly Inelastic Demand Curve

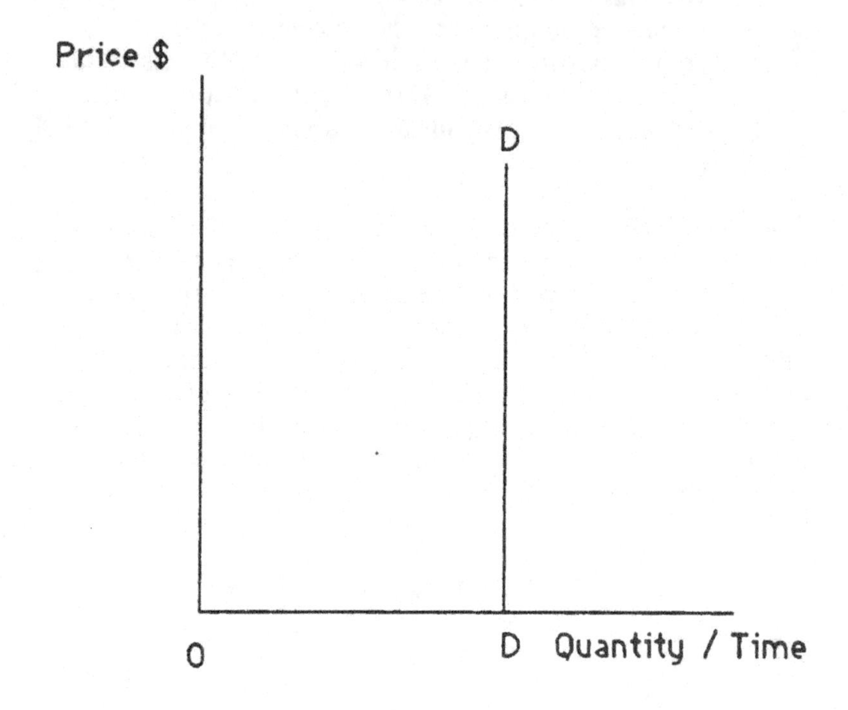

Price elasticity is from zero to infinity. If the absolute value (omitting the sign) of the coefficient is greater than 1, demand is said to be *elastic*. That is, the percentage change in quantity demanded is greater than the percentage change in price. Quantity is said to be responsive to price changes. If the absolute value is less than 1, demand is *inelastic*. The percentage change in quantity is less than the corresponding percentage change in price. Quantity demanded is relatively unresponsive to price changes. A coefficient of 1 represents the case of unitary elasticity. The percentage change in quantity equals the percentage change in price.

The elasticity coefficient varies along the demand curve for most functional forms of the curve. In Figure 3-7, we see that at the point where the demand curve approaches the Y-axis, we have infinite elasticity. At the center of the demand curve we have unitary elasticity and at the point where the demand curve touches the X-axis we have zero elasticity. Note that slope is constant in Figure 3-7, but elasticity varies throughout the curve.

Calculation of Own Price Elasticity of Demand

From Table 3-1 earlier in this chapter, suppose there is a change in price (ΔP) from $5.00 to $4.00; the resulting change in quantity demanded will be ΔQ. If the change in price (ΔP) is very small, one may compute the point elasticity of demand:

$$E_p = \frac{\frac{\Delta Q_D}{Q_D}}{\frac{\Delta P}{P}}$$

$$E_p = \frac{\frac{5-10}{5}}{\frac{5.00-4.00}{5.00}} = \frac{-5}{5} \times \frac{5}{1}$$

$= -5$, or $|5|$ since we are concerned with absolute values.

The price elasticity is -5, which means that a 10 percent increase in price will result in a 50 percent decrease in quantity demanded.

Consider another example, from Table 3-2. Suppose price falls from $1.00 to $0.50, quantity demanded rises from 100,000 to 300,000, and P * Q or total expenditure rises to $150,000. Then price elasticity of demand is:

$$E_p = \frac{\frac{\Delta Q}{Q}}{\frac{\Delta P}{P}} = \frac{(100{,}000-300{,}000)/100{,}000}{(\$1.00-\$0.50)/\$1.00} = -\frac{2}{.5} = -4$$

The coefficient is greater than 1. But some caution must be exercised. Q and P are definitely known, but suppose price *rises* from

\$.50 to \$1.00, quantity changes from 300,000 to 100,000. Price changes by \$.50. Now try the other values of P and Q.

$$E_p = \frac{(300{,}000 - 100{,}000)/300{,}000}{(\$.50 - \$1.00)/(\$.50)} = \frac{.67}{-1.0} = -.67$$

Demand appears to be inelastic.

The difficulty is that elasticity has been computed over a wide arc of the demand curve, but evaluated at a specific point. We can get a much better approximation by using the average values of P and Q over the arc. That is, for large changes such as this we should compute E_p as:

$$E_p = \frac{\dfrac{Q_1 - Q_0}{Q_1 + Q_0}}{\dfrac{P_1 - P_0}{P_1 + P_0}}$$

where subscripts 1 and 0 refer to new and to initial prices and quantities demanded. Using this formula, obtain

$$E_p = \frac{(100{,}000 - 300{,}000)/100{,}000 + 300{,}000)}{(\$1.00 - \$0.50)/(\$1.00 + \$0.50)} = -\frac{3}{2}$$

Demand appears to be elastic.

These differences are not a major problem when elasticities are calculated empirically since the demand curve is usually represented by an algebraic equation. Thus, the demand equation may be represented as: $Q = F(P_1, P_2, Y_1)$, where: P_1 is the price of the good and P_2 the price of a substitute good, and Y refers to income level. Therefore, in calculating the point elasticity of demand use the derivative form:

$$E_p = \frac{dQ}{dP} * \frac{P}{Q}$$

dQ represents a minute change.

The slope of the demand curve is represented by $\frac{dQ}{dP}$. Assume now that the demand curve for shrimp in country X is represented by the equation:

$$Q_{d_s} = -3{,}000\ P_s + 1{,}000\ Y + 0.05\ Pop + 1{,}500\ P_c$$

where:

Q_{d_s} = the average quantity of shrimp in kg. purchased in 1986
P_s = the average market price per kg. of shrimp in French francs
Y = per capita disposable income in country X in French francs
Pop = population in thousands
P_c = the average price in French francs per kg. of catfish in lbs. in 1986.
(Conversion rate: 1 kg. = 2.2 lbs.)

The change in the quantity of shrimp purchased resulting from a change in price is the partial derivative:

$$\frac{dQ_s}{dP_s} = -3{,}000$$

Using the formula for price elasticity of demand:

$$E_p = \frac{dQ_{d_s}}{dP_s} * \frac{P_s}{Q_{d_s}}$$

and calculating the elasticity at two points on a hypothetical demand curve:
(1) where P_1 = \$3,000 and Q_1 = 9,500,000 and (2) where P_2 = \$3,500 and Q_2 = 8,000,000 we obtain:

$$(1)\ E_p = (-3{,}000)\ \frac{3{,}000}{9{,}500{,}000} = -0.95$$

$$(2)\ E_p = (-3{,}000)\ \frac{3{,}500}{8{,}000{,}000} = -1.3$$

At the first point, a 1 percent change in the price of shrimp will result in a 0.95 percent change in the quantity demanded (inelastic). At the second point on the demand curve, a 1 percent change results in a 1.3 percent change in demand (elastic).

Determinants of Price Elasticity

There are three major causes for differential price elasticities:

1. the extent to which a good is considered to be a necessity;
2. the availability of substitute goods; and
3. the proportion of income spent on the product.

If the good is considered a basic necessity, the price elasticity of demand will be very inelastic. That is, a 1 percent change in price will result in less than a 1 percent change in quantity demanded. The availability of substitute goods will increase the price elasticity of demand. If price is increased, all other factors remaining constant, and substitutes are available, consumers will switch from the relatively high priced good to the relatively low priced good. The larger the proportion of the income spent on the good the greater the price elasticity of demand. The price elasticity of demand for fish varies with the type of fish and the form in which the product is presented.

Fish are considered a necessity by consumers in developing countries, hence, the demand is inelastic. The research conducted on seafood shows that seafood consumption is not highly related to red meat and poultry consumption. Thus, the substitution effect is weak.

FIGURE 3-5. A Sloping Demand Curve Relatively Elastic

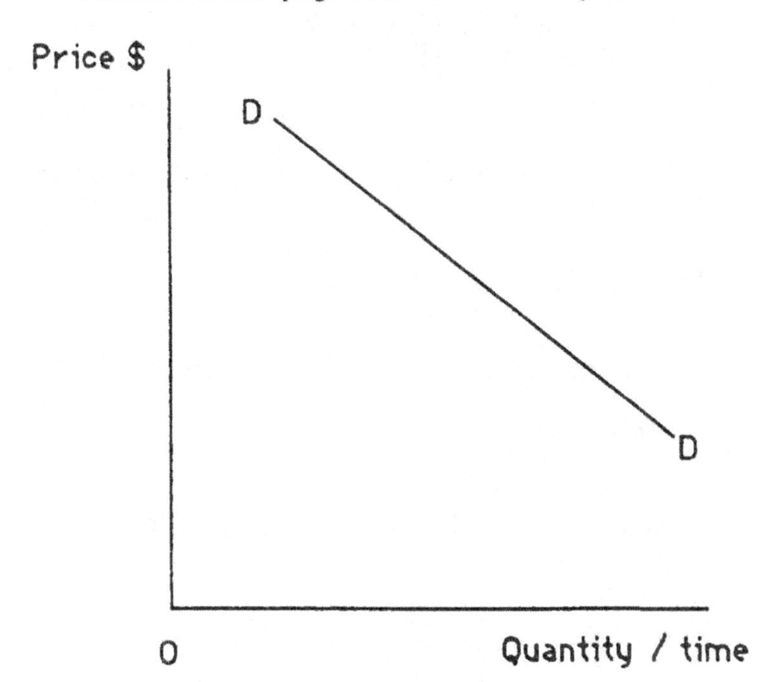

Income Elasticity

Income elasticity of demand is a measure of the responsiveness of quantity purchased to changes in income, other factors held constant. The income elasticity may be represented by:

$$E_y = \frac{\dfrac{\Delta Q_{d_s}}{Q_{d_s}}}{\dfrac{\Delta Y}{Y}} = \frac{\Delta Q_{d_s}}{Q_{d_s}} * \frac{Y}{\Delta Y}$$

where:

Y = income; Q = quantity demanded.

This equation measures the percentage change in quantity corresponding to a 1 percent change in income, other factors held constant. From our previous equation:

$$Q_{d_s} = -3{,}000\ P_s + 1{,}000\ Y + 0.05\ Pop + 1.500\ P_c$$

the quantity change resulting from an income change is the partial derivative:

$$\frac{dQ_{d_s}}{dY} = +1{,}000$$

The formula for the income elasticity of demand is:

$$E_y = \frac{dQ_s}{dY} * \frac{Y}{Q}$$

The income elasticity of demand from point Q = 8,000,000 and income (Y) = 10,000 is:

$$E_y = +\ 1{,}000\ \frac{10{,}000}{8{,}000{,}000} = +\ 1.25$$

This means that a 1.0 percent change in income will result in a 1.25 percent change in quantity demanded (income elastic). In most cases, the coefficient is positive since it is expected that as income increases consumers will buy more.

A negative income elasticity denotes that the consumer will buy less and less of the good as income increases. The good is then said to be inferior. In many West African countries, catfish caught in rivers and streams are considered inferior. Only low income consumers purchase catfish and as consumer income increases, consumers purchase other species such as mullet and tilapia. The income elasticity for catfish in these areas is expected to be negative. In the United States, income elasticity for catfish was found to be negative by Dellenbarger et al. A good with an income elasticity between 0 and 1 is said to be normal, and a good with an income elasticity greater than 1 is said to be superior. Shrimp is considered a superior good by consumers of every strata; the income elasticity for shrimp is usually greater than 1.

Cross Price Elasticity

Cross price elasticity measures the responsiveness of quantity demanded of one good to changes in the price of a related good, every other price remaining constant. The cross price elasticity of shrimp resulting from changes in catfish price is:

$$E_{sc} = \frac{\frac{\Delta Q_s}{Q_s}}{\frac{\Delta P_c}{P_c}} = \frac{\Delta Q_s}{\Delta P_c} * \frac{P_c}{Q_s}$$

where:

Q_s = quantity of shrimp purchased
PC = price of catfish

From the equation on pages 47 and 50:

$$Q_{d_s} = -3{,}000\ P_s + 1{,}000\ Y + 0.05\ Pop + 1{,}500\ P_c$$

the cross price elasticity for shrimp where Q_s = 9,500,000 and Pc = 3,500 is:

$$E_{sc} = (+\ 1500) * \frac{3{,}500}{9{,}500{,}000}$$

$$E_{sc} = +\ 1500 * \frac{3{,}500}{9{,}500{,}000}$$

$$E_{sc} = +.55$$

This means a 1 percent increase in the price of catfish will result in a 0.55 percent change in the quantity demanded of shrimp. Therefore, catfish and shrimp may be considered as substitutes in this case.

There are three types of cross relationships: the commodities might be substitutes, complements, or independent. The substitution effect (cross price elasticity) is positive for substitute commodities and negative for complementary commodities. Substitution effects are 0 for independent commodities.

FIGURE 3-6. A Sloping Demand Curve Relatively Inelastic

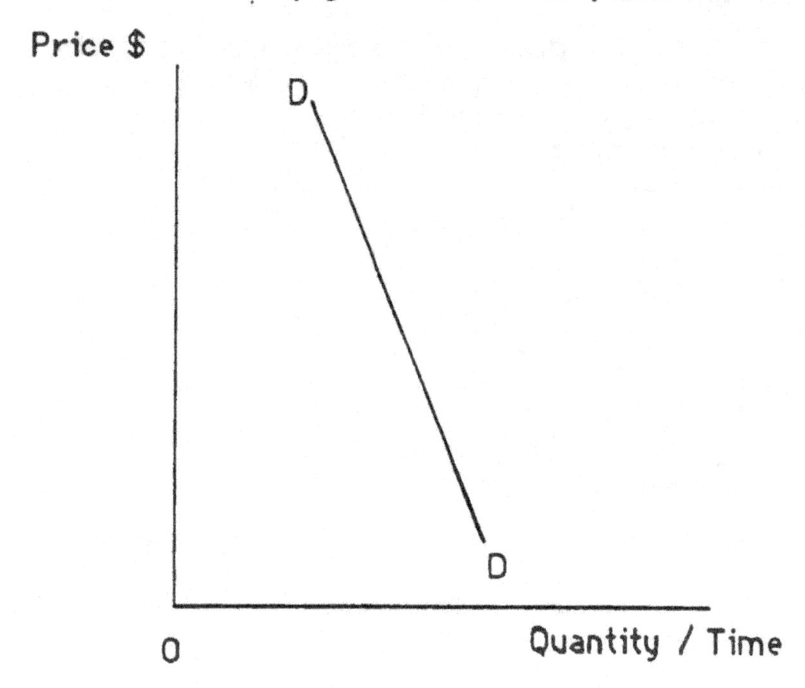

An Applied Demand Curve for Catfish

The quantity demanded of processed catfish at the wholesale level for the United States was estimated, using time series monthly data from January, 1980 to December, 1989. The model proposed was:

$$QD_C = f(P_c, Y, I)$$

where:

QD_C = per 1000 capita of fish purchased in lbs.
P_c = real average wholesale price of processed catfish in dollars
Y = real disposable income per 1000 capita from 1980 to 1989
I = per 1000 capita of processed imported catfish.

The model obtained was:

$$QD_C = -61.94 - 23.92P_c + 10.58Y + 0.35I$$
$$(-3.30) \quad (-4.17)^* \quad (11.07)^* \quad (0.06)$$
$$R^2 = .73$$

* significant at .05 probability level

Model results indicated that quantity demanded is inversely related to price. The price elasticity of demand at the mean was −1.08, which means that a 1.0 percent increase in price will result in a 1.08 percent change in demand. Price is said to be elastic in this case. The model also showed that demand per capita is positively

FIGURE 3-7. Elasticities Along a Demand Curve

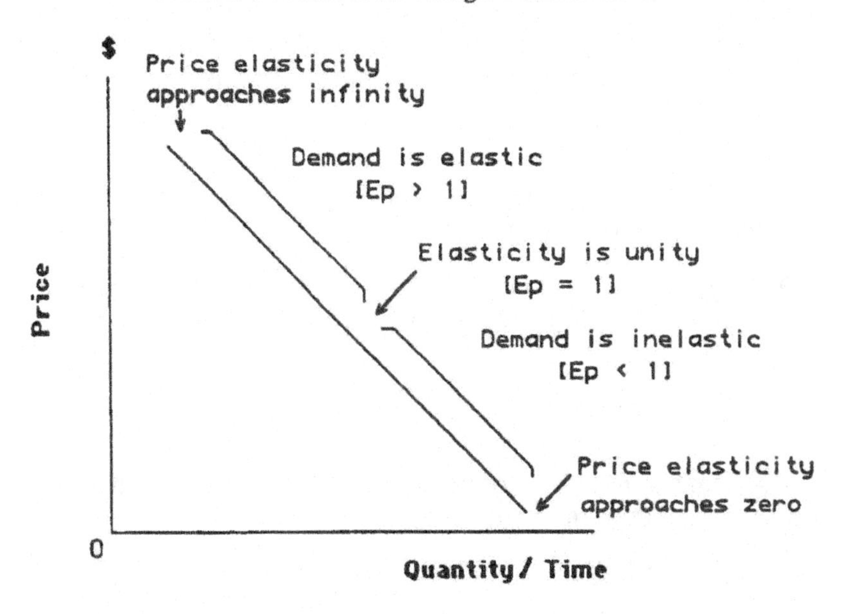

TABLE 3-2. Hypothetical Quantity Demanded and Expenditure for Fish at Various Prices

Price	Quantity demanded	Total expenditure
1.00	100,000	100,000
.50	300,000	150,000
.25	600,000	150,000
.10	1,000,000	100,000

related to disposable income. The income elasticity of demand was 3.66 at the mean, which indicates that processed catfish may be considered a superior good. A 1 percent change in income will result in a 3.7 percent change in quantity demanded of processed catfish.

ELASTICITY, TOTAL AND MARGINAL REVENUE

Throughout the remainder of this book, concern will be focused on revenue derived from fish farming. Thus, a brief review of the way revenue is measured is in order. Since revenue is related to demand, a further digression of elasticity and demand is also in order.

The total revenue (TR) of a fish farm is the total sales over a specific time period, assuming that the fish farmer is engaged only in the production and sale of fish. If all the fish produced are sold at the same price, the total revenue is derived by multiplying price (P) by the total quantity sold (Q):

$$TR = P \times Q$$

Average and Marginal Revenue

Average revenue (AR) is the amount of money received per unit sold. It is equal to total revenue divided by the quantity sold:

$$AR = \frac{TR}{Q}$$

Marginal revenue (MR) is the rate of change of total revenue with respect to a change in the quantity sold. It is calculated:

$$MR = \frac{\Delta TR}{\Delta Q}$$

or, said another way: MR is the first derivative of TR.

A look at Table 3-3 shows the calculated TR, AR, and MR for quantities of dressed catfish sold during a given time period. The table shows annual total revenue, average revenue, marginal revenue, and price elasticities. The astute student should be able to sketch the relationship of these revenue measures in graph form. What will MR be when TR is at its maximum point?

We see that the absolute values of the elasticities were decreasing as quantity purchased increased and prices declined. The calculations of elasticity were made over an arc. Therefore the formula:

$$E_p = \frac{Q_0 - Q_1}{Q_0 + Q_1} \times \frac{P_0 + P_1}{P_0 - P_1}$$

was used to calculate the elasticities. The elasticity is 1.0 somewhere between TR equal to 1250.00 and 1200.00. That is, the absolute value of the elasticity is 1.0 where TR is maximum. The absolute value of the elasticity is less than 1.0 after TR is maximum and MR is less than 0.

According to Figure 3-8, in the elastic portion of the demand curve, total revenue is increasing. It reaches the maximum at unitary elasticity, then declines in the inelastic portion of the demand curve. Marginal revenue is greater than 0, but declining in the elastic portion of the demand curve. It reaches 0 at the point of unitary elasticity, and is less than 0 in the inelastic portion of the demand curve.

What does this mean to the fish culturist? If the demand is elastic, a decrease in the price of fish will result in proportionately greater expansion of quantity demanded. An increase in the price of fish will result in a proportionately greater decline in quantity demanded. Hence, a marketing strategy based on price may be deter-

TABLE 3-3. Total Annual Revenue, Average Revenue, Marginal Revenue and Price Elasticity for Dressed Catfish Sales, (Hypothetical Data)

Quantity/ Q Kg	Price P $	Total revenue TR = P x Q $	Average revenue AR = TR/Q = P	Marginal revenue MR = ΔTR/ΔQ	Elasticity
0	5.00	0	-	-	-
100	4.50	450.00	4.50	4.50	-19.00
200	4.00	800.00	4.00	3.50	- 5.66
300	3.50	1,050.00	3.50	2.50	- 3.00
400	3.00	1,200.00	3.00	1.50	- 1.86
500	2.50	1,250.00	2.50	.50	- 1.22
600	2.00	1,200.00	2.00	-.50	- .82

FIGURE 3-8. Relations Among Price Elasticity, Marginal Revenue and Total Revenue

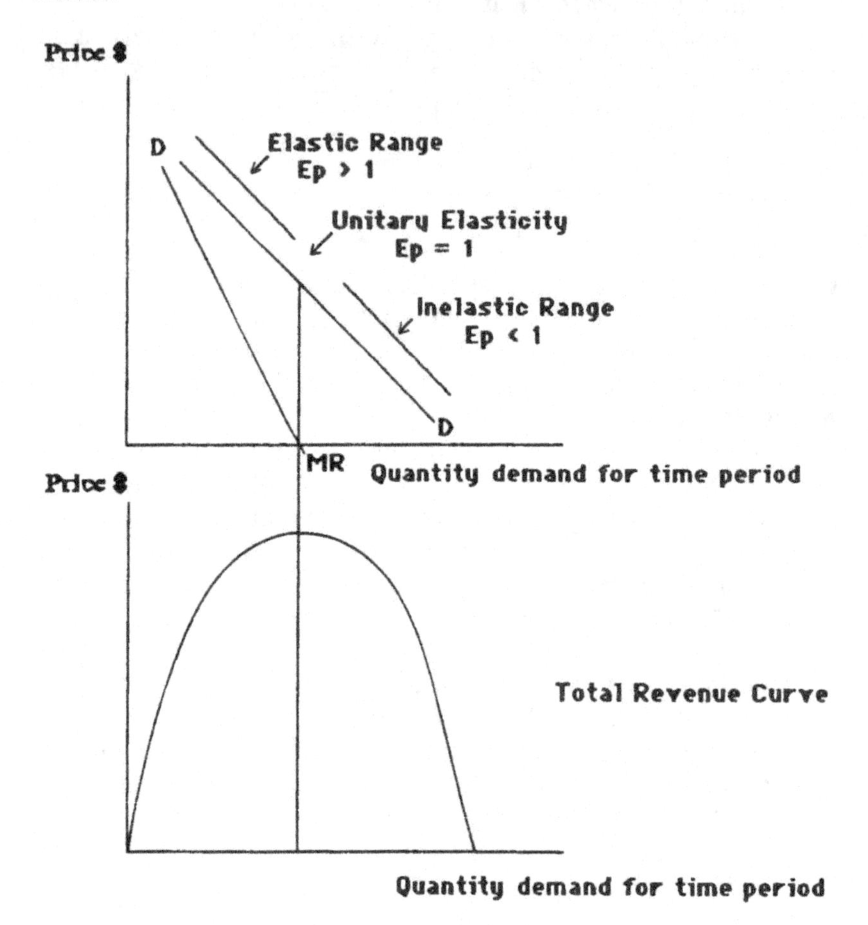

mined. Therefore, if demand is elastic, a decrease in price will increase total revenue because quantity demanded increases proportionately more than price declines, and a price increase will result in less total revenue because quantity demanded falls proportionately more than price increases. The reverse is true for an inelastic demand curve. An increase in prices or a reduction in quantity supplied will result in an increase in total revenue.

Market Demand

Market demand is defined in terms of the alternative quantities of a commodity which all consumers in a particular market are willing and able to buy as price varies and as all other factors are held constant. As indicated earlier, a market demand curve may be thought of as a summation of individual demand relations.

If there are two individual purchasers of a good, the market demand curve may be expressed as:

$$q_t = q_A + q_B$$

where:

q_t = total quantity demanded by individuals A and B
q_A = quantity demanded by individual A
q_B = quantity demanded by individual B.

It is most often true that A and B will be confronted with the same relative price of the good.

DERIVED DEMAND

Derived demand is the demand for a product that occurs due to a demand for a primary product. For example, the demand for catfish at the processing plant is derived from products such as fillet, fish nuggets, and fish cakes produced from catfish. The demand for land, labor, and capital used to produce fish is derived from the demand for fish products.

In fish market studies it is sometimes difficult to directly estimate price elasticities of demand for a product at different points in the marketing chain. Also, the elasticities are likely to be different at the retail and farm level. It is possible, however, with knowledge of the marketing margins and of the elasticity at one level (say at the retail) to estimate the elasticity at another level (at the farm).

The exact relationship between elasticities depends on how primary and derived demand curves are related. Since the two curves are separated by a schedule of marketing margins, the problem is how marketing margins behave.

In this case there are two parallel demand curves which means the margin is constant regardless of the amount marketed. Elasticity is calculated using the formula:

$$E_d = E_r \left(\frac{P_d}{P_r}\right)$$

where:
P_d = price at derived demand at farm level
P_r = price at primary (market) demand

If the margins decrease with lower prices as the quantity marketed increases, the derived demand elasticity is now:

$$E_d = E_r \left[1 - \frac{a}{(1 - b)P_r}\right]$$

since $M = a + b\,P_r$

where:

M = margin
a = constant
b = coefficient of regression.

The derived demand elasticity is less elastic, or more inelastic, than the related demand curve.

PRODUCER SUPPLY

The supply of fish and shellfish for human utilization may be expanded through discovery of new resources, exploitation of known, but untapped resources, introduction of improved fishing technology, increased emphasis on cultivation, and better handling of catches to reduce losses through spoilage. The supply of aquacultural products may be increased by increasing the resources allocated to fish production and improving the technology of fish production. Production and technology could, in turn, be enhanced by the increase in the real prices for the various species.

Supply may be defined in terms analogous to demand. Supply is a schedule of alternative quantities of a good or service offered for sale at different prices. It is the amount of goods and services that producers are willing and able to offer in the marketplace at specific prices. Supply may be represented by a schedule of quantities and prices, or graphically by a curve, or by an algebraic function.

Table 3-4 shows how a typical firm might behave when supplying goods and services at varying prices. The firm is willing and able to sell more at a higher price within certain limits.

The supply schedule for tilapia shows that quantity and price increase or decrease in the same direction. Figure 3-9 shows a supply curve for tilapia which corresponds to Table 3-4. The supply curve has a positive slope which indicates that more of a commodity will be placed on the market when its relative price increases, all other variables remaining constant. Note that the curve in Figure 3-9 is "fit" to the data using the regression shown rather than drawn through the precise points shown in Table 3-4. This procedure allows viewing a smooth curve of supply through time.

Quantity Supplied and Supply

As in the case of demand, *quantity supplied* is a static concept. It shows movement along a curve as quantities and prices vary. *Supply* is a more dynamic concept. Supply changes reveal a shift in the supply curve from right to left or left to right. Figure 3-10 shows a shift in the supply curve from right to left (S_0S_0 to S_1S_1) which is a decrease in supply. A shift of the curve from S_0S_0 to S_2S_2 signifies an increase in supply.

TABLE 3-4. Hypothetical Supply Schedule for Tilapia

Price/Kg. $	Quantity (1,000 kg.)
4.00	250
3.50	190
3.00	140
2.00	100
1.50	0

FIGURE 3-9. Hypothetical Supply Curve for Tilapia

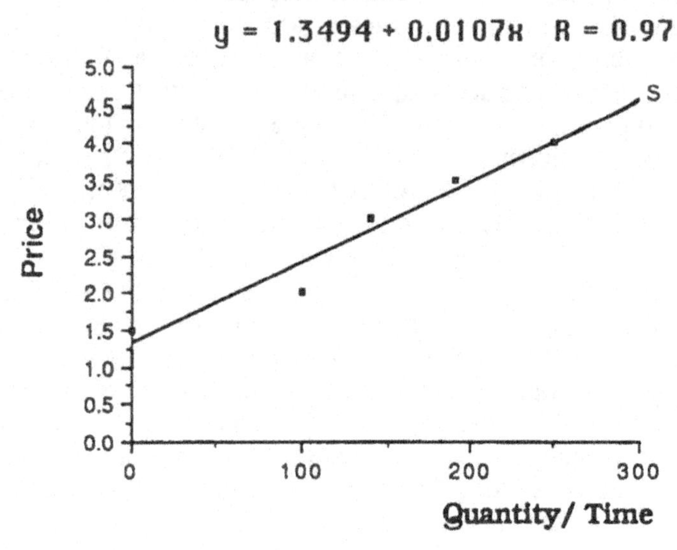

Factors Influencing the Supply of Fish and Fish Products

Factors may be grouped in five major categories:

1. changes in the price of inputs (factors of production);
2. changes in profitability of substitute commodities;
3. changes in production and marketing technology;
4. changes in prices of joint products (commodities that are produced together); and
5. institutional and environmental changes such as government programs.

Changes in Price of Inputs

An increase in the price of feed, seed, and water may result in a decrease in the amount of fish supplied to the market, all things remaining constant. A drop in the price of inputs may result in the increase in the supply of fish, all other things remaining constant. A fall in the price of fingerlings and feed have positive effects on the quantities of catfish supplied. An increase in pond cost, on the other hand, has a negative effect on the quantity of fish supplied.

FIGURE 3-10. Shifts in the Supply Curve

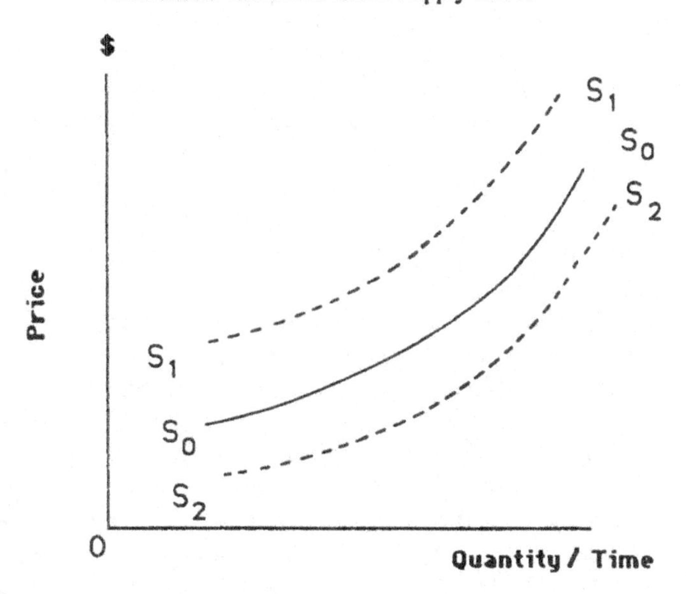

Changes in Profitability of Substitute Commodities

Farmers are likely to allocate more resources to the competing enterprise which brings the higher profit, all things remaining constant. In the United States Mississippi River delta, farmers have been increasing the land area devoted to catfish while decreasing lands placed in rice and cotton production. In Jamaica, where sugarcane has been producing only marginal returns, farmers have been reallocating sugarcane lands to tilapia production.

Changes in Production and Marketing Technology

Changes in technology have brought rapid gains to the aquacultural industry. For example, developments in producing fingerlings have resulted in each farmer producing on-farm stock instead of depending on a central source. Also, fish breeding has given rise to production of all male tilapia which allows more rapid growth of tilapia to marketable size.

In the 1950s, world aquaculture produced less than 20,000 metric

(20,320 U.S.) tons of tilapia a year. By 1983, 10,000 more metric (10,160 U.S.) tons were being produced, primarily in Asia. About 20 years ago, the Java tilapia (*Oreochromis mossambicus*) became recognized internationally, but its appearance and yields were poor. The Nile tilapia (*Oreochromis niloticus*) soon proved superior, but even faster growing hybrid fish were obtained when the two species were crossed. The Nile tilapia was then crossed with the blue tilapia (*Oreochromis aureus*) to yield only male hybrids, which boosted growth even more. Finally, researchers crossed Nile tilapia with albino Java tilapia and obtained an extremely fast-growing red variety. This hybrid was commercially produced in the late 1970s. Subsequently, tilapia production rose a hundredfold.

New methods of harvesting, processing, preserving, and storing fish have also resulted in increased supply of fish. The use of tractors in hauling nets reduced the difficulties in harvesting fish, especially when the water is cold. Improvements in processing fish, such as smoking, canning, pre-packing, irradiation, and freeze drying, increased the quantity of fish available at any one time. New developments in biotechnology may shift the supply curve further to the right as more rapid rates of growth per time period result from genetic engineering.

Changes in Prices of Joint Products

Changes in the prices of joint products will either have a positive or negative effect on the supply of fish. For example, an increase in the price of fish eggs will have a positive effect on the price of the brood fish. Pearl culture results in production of mother-of-pearl shell and oyster meat. The black lip mother-of-pearl oyster, *Pinctada magaritifera*, has long been prized and collected for its high quality nacre. An increase in the price of pearls will affect the quantity of oysters produced, and an increase in the demand for oyster meat will affect the quantity of mother-of-pearl produced.

Institutional and Environmental Changes

Institutional and environmental problems may have either positive or negative effects on fish supply. Government programs such as credit availability, subsidies, and infrastructure development likewise impact the fish supply. Government regulations limit the species that may be introduced into a country, thereby affecting potential as well as realized supplies. If new high quality and highly productive species are not allowed to be introduced, existing stocks may be depleted more rapidly. On the other hand, introducing exotic species has been detrimental to native stocks in many instances. Environmental factors such as diseases and water pollution generally have negative impacts on fish supply by reducing quality or fish stocks.

ELASTICITY OF SUPPLY

Price Elasticity of Supply

The price elasticity of supply expresses the percentage change in quantity supplied in response to a 1 percent change in price, other factors held constant. In algebraic terms, it is expressed as follows:

$$E_s = \frac{\frac{\Delta Q}{Q}}{\frac{\Delta P}{P}} = \frac{\Delta Q}{\Delta P} * \frac{P}{Q}$$

where:

Q = quantity supplied
P = price.

Zero elasticity means that supply is fixed. In that case supply is said to be perfectly inelastic and the supply curve would be a vertical line at the point of total supply. There is no change in quantity irrespective of price changes. A relatively inelastic supply refers to the range of elasticities between 0 and 1. Quantity supplied is *relatively* unresponsive to price changes.

An elastic supply refers to a situation in which the supply elasticity coefficient is greater than 1. That means a 1 percent change in price will result in a percentage change in quantity supplied greater than 1. Figure 3-11 illustrates supply curves with three different levels of elasticities.

For all lines going through the origin, $E_s = 1$. For lines that cut the Y-axis (price axis) at a positive value, $E_s > 1$. Lines that cut the Y-axis (price axis) at a negative value or positive X-axis (quantity axis) have $E_s < 1$.

Calculating Supply Elasticities

If ΔP is not too small, the arc elasticity of supply may be calculated by using a formula similar to that shown in the discussion on demand. That is:

$$E_s = \frac{\Delta Q}{Q_1 + Q_2} \div \frac{\Delta P}{P_1 + P_2}$$

$$E_s = \frac{\Delta Q}{\Delta P} \times \frac{P_1 + P_2}{Q_1 + Q_2}$$

FIGURE 3-11. Supply Curves Showing Different Elasticities

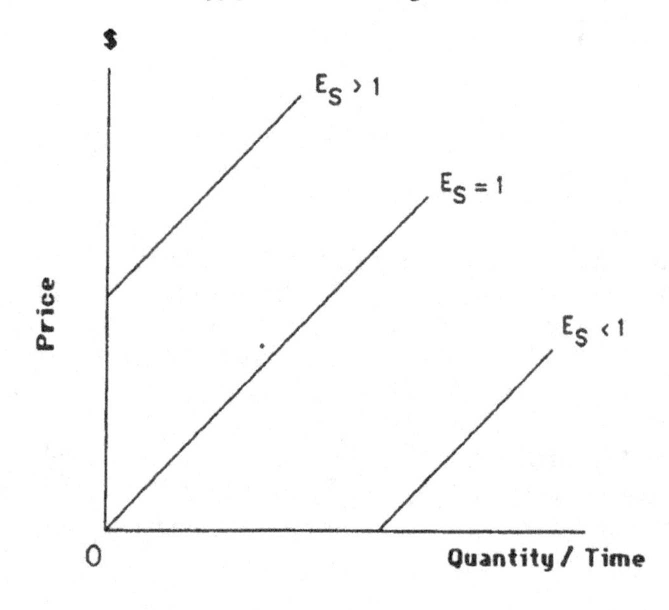

Suppose one would like to calculate the price elasticity of supply for tilapia from Table 3-4 when price changes from \$4.00 to \$3.50; the arc elasticity of supply is:

$$E_s = \frac{(250 - 190)}{(4.00 - 3.50)} * \frac{(4.00 + 3.50)}{(250 + 190)}$$

$$E_s = \frac{60}{.50} * \frac{7.50}{440} = 2.04$$

$$E_s = 2.04$$

Therefore, a 1.0 percent change in price will result in a 2.04 percent change in quantity supplied. Supply is said to be elastic.

The supply curve is usually represented by an algebraic equation such as:

$$Q_s = f\,(P, P_2, \ldots)$$

where:

Q_s = quantity supplied
P_1 = price of fish
P_2 = price of factors of production.

Therefore, the elasticity of supply for a small change in price is:

$$E_s = \frac{dQ}{dP} * \frac{P}{Q_s}$$

Given the theoretical supply function for catfish:

$$Q_c = 30 + 5\,P_c - 2\,P_f$$

where:

Q_c = the quantity of catfish produced in tons in a given year
P_c = price of catfish dollars per ton
P_f = price of feed dollars per ton.

The price elasticity of supply at $P_c = 4$ and $Q_c = 30$ is computed using the formula:

$$E_s = \frac{dQ}{dP} * \frac{P}{Q_s}$$

and we find

$$E_s = 5 * \frac{4}{30} = .66$$

This means the supply function of catfish is relatively inelastic at this point; that is, a 1.0 percent change in price will result in a 0.66 percent change in quantity supplied.

The supply elasticity due to a change in the price of feed when $Q_c = 30$ and $P_f = 20$ is:

$$E_{s_f} = \frac{dQ_c}{dP_f} * \frac{P}{Q_c}$$

$$= -2 * \frac{20}{30} = -1.33$$

Thus, an increase in the price of feed has a negative effect on the quantity supplied of catfish. A 1.0 percent increase in the price of catfish will result in a 1.33 percent decrease in the quantity of catfish supplied.

An Applied Supply Curve for Catfish

A supply curve was determined using time series data collected from January 1980 to January 1989. The model was expressed as:

$$QS_c = f(P_c, P_f, IM)$$

where:

QS_c = quantity sold in lbs. of processed catfish, 1980 to 1989

P_c = real farm price of catfish in dollars, 1980 to 1989

P_f = the price of feed in dollars per lb., 1980 to 1989,

IM = per 1000 capita of imported processed catfish.

The results obtained were:

$$QS_c = 31.48 + 25.41P_c - 0.52P_f + 0.39IM$$
$$(1.86)^* \quad (1.82)^* \quad (0.82)$$
$$R^2 = .54$$

* significant at the .10 probability level

These data show that the quantity supplied is positively related to the price of fish.

Price Flexibilities

Price flexibility is usually treated as the inverse of price elasticity. The flexibility coefficient gives the percentage change in price associated with 1 percentage change in quantity, other factors held constant. The price flexibility may be calculated:

$$F_p = \frac{\frac{\Delta P}{P}}{\frac{\Delta Q}{Q}} = \frac{\Delta P}{\Delta Q} * \frac{Q}{P}$$

The flexibility coefficient implies that price is a function of the quantity of a particular product as well as the quantities of substitutes. Price flexibility is common in ocean fishery supply studies where it indicates the responsiveness of changes in price to changes in quantity supplied.

Short- and Long-Run Supply Curves

Generally, the longer the time period involved, the greater the degree of elasticity of supply. Alfred Marshall, a distinguished English economist, illustrated three different types of supply curves according to the time period involved. His example dealt with the London fish market. In the very short-run, or market period, the supply is relatively fixed. The boats could go out only one time per day and, hence, could not adjust the quantity offered in response to price. This result leaves a supply elasticity of approximately 0. In the short-run, producers could respond more to price changes by sending out more (or less) workers within the limits of their fixed (in the short-run) complement of equipment (see Figure 3-12).

Therefore, a higher degree of elasticity or responsiveness to price changes would be expected here than in the market period. In the long-run, a still higher degree of elasticity would be expected since the firm would have the ability to construct new capital equipment such as boats, nets, and even docks and buildings. Hence considerable opportunity to adjust would exist.

COMPETITIVE MARKET EQUILIBRIUM

Equilibrium in the market is a state which depends on and satisfies the existing conditions of demand and supply at any given moment of time. It implies a situation of rest, or absence of change, over a period of time. Figure 3-13 illustrates a simple equilibrium in a market in which supply and demand are equal.

At price P_e, sellers wish to sell exactly the quantity which buyers wish to buy. The market clears at this price. If the price were raised

FIGURE 3-12. Supply Curves Showing Elasticity Relative to Time

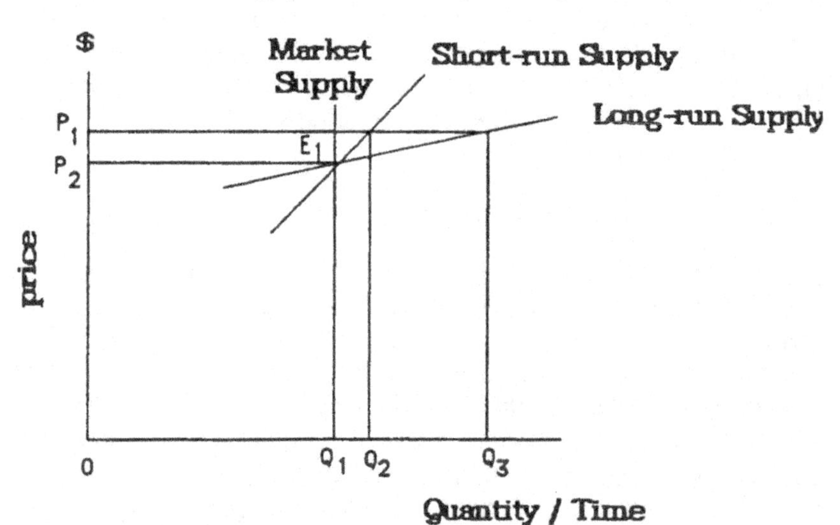

FIGURE 3-13. An Illustration of Competitive Market Equilibrium

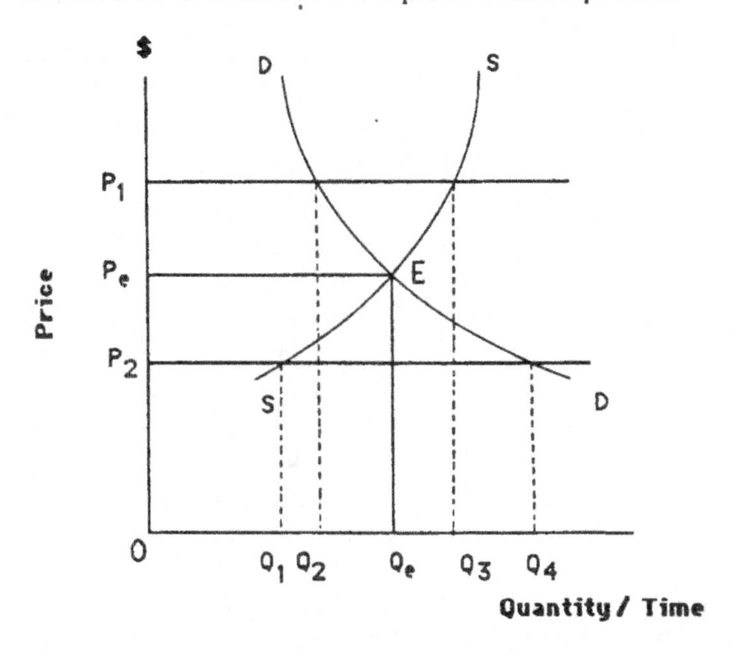

by some means to P_1 there would be a contrived surplus. Suppliers of fish would be willing to supply quantity Q_3 while fish consumers would be willing to purchase quantity Q_2. The contrived surplus would be quantity Q_2Q_3. Conversely, if the price were lowered by some means to P_2 there would be a contrived shortage. Fish consumers would be willing to purchase Q_4, but fish producers would be ready to supply only quantity Q_1. There would be a contrived shortage quantity Q_1Q_4. In a free market, that is, one in which no outside force interferes, the market will tend to establish the price at equilibrium level. Governments may interfere with various programs which might have serious effects on prices and quantities in the marketplace.

A few examples illustrate the interrelationship of demand and supply. With a given demand curve, an increase in supply tends to lower the equilibrium price. In Figure 3-14, an illustration of an increase in supply is given in which the price decreases from P_1 to P_2 in an area of inelastic demand. The total revenue to the seller is

FIGURE 3-14. Supply Increase in an Inelastic Demand Range

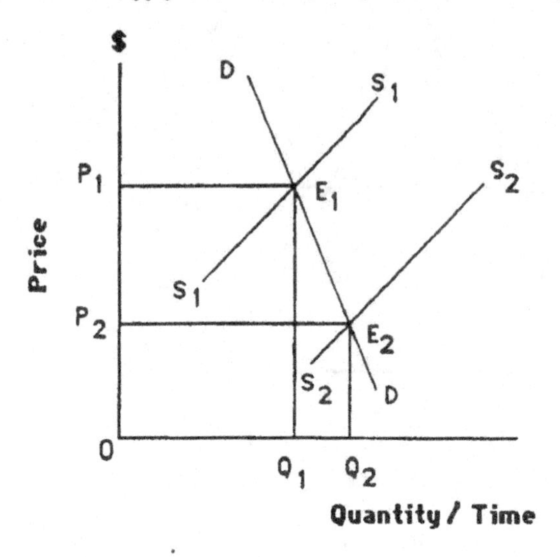

drastically reduced, being measured by the rectangle 0, P_2, E_2, Q_2. Before the increase in supply, the total revenue was P_1, E_1, Q_1, 0. It is somewhat ironic that when nature cooperates by giving a bountiful supply of some crop, it may be a disaster to the seller in terms of total revenue. (Note: Remember, a change in supply is a movement of the entire supply curve. A change in quantity supplied is a movement along a given supply curve.)

In developing countries, the demand for certain fish species such as tilapia is inelastic for a given segment of the population. If it is assumed that farmers are aware of the nature of demand elasticity, would farmers be willing to adopt a new technology that would then double the supply of tilapia? Let's assume that a young extension worker has learned from a nutritionist that there is a new type of protein feed that would shift the supply curve to the right. From Figure 3-14, the supply curve S_1S_1 would be shifted to S_2S_2. Without knowing anything about the cost of feed, it is apparent from the figure that farmers total revenue would be reduced. If the farmer were using compost to fertilize a pond (assuming the price of compost is 0), but was advised to use the new feed that would increase the supply within a given time period, the farmer should be very reluctant to adopt this new technology, unless the new feed was cost-reducing and the demand curve was shifted significantly to the right. On the other hand, if demand were elastic, or if supply shifts within the elastic portion of the curve, as in Figure 3-15, the farmer's total revenue is likely to increase from 0, Q_1, E_1, P_1, to 0, Q_2, E_2, P_2. This phenomenon helps to explain why farmers are more likely to adopt new technology when producing fish with high price elasticity of demand than species with low price elasticity.

If supply is highly elastic, a given increase in demand may only affect price by a small amount, as shown in Figure 3-16. If supply is highly inelastic and demand increases, the magnitude of the price increase will be quite noticeable (see Figure 3-17). The increase in total revenue shown in Figure 3-17 may be significant. Governments of developing countries are very quick to regulate the price of species with inelastic supplies, especially when a large portion of the population is dependent on this species as a protein source.

FIGURE 3-15. Supply Increase in Elastic Demand Range

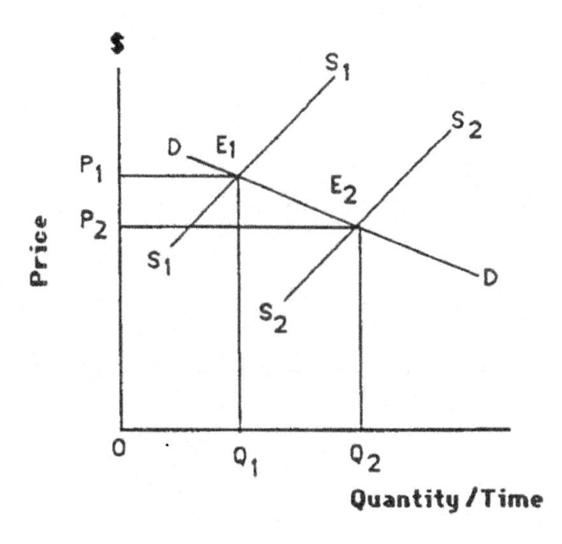

FIGURE 3-16. Price Change with Demand Increase: Elastic Supply

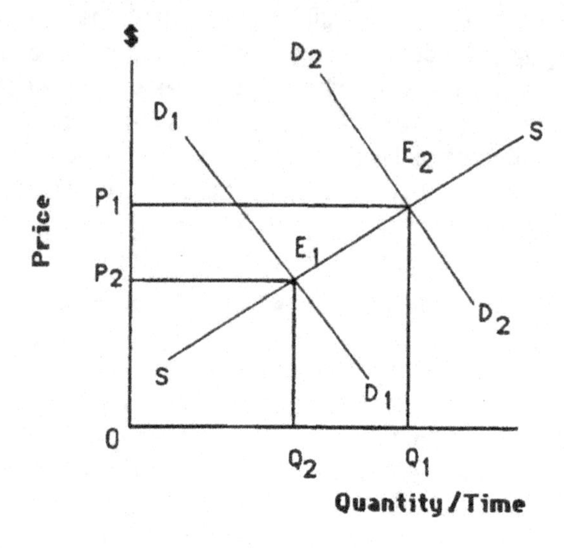

FIGURE 3-17. Price Change with Demand Increase: Inelastic Supply

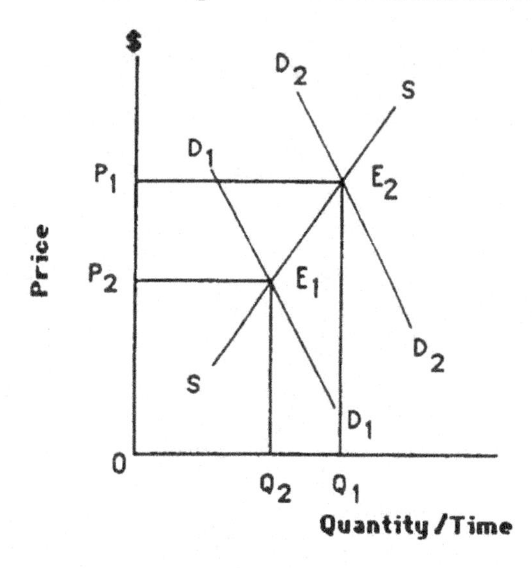

REFERENCES AND RECOMMENDED READINGS

Albrecht, W.P., Jr. 1983. *Economics*, Fourth Ed. Englewood Cliffs, NJ: Prentice-Hall.

Armstrong, M.S., C.E. Boyd, and R.T. Lovell. 1987. "Environmental Factors Affecting Flavor of Channel Catfish from Production Ponds." *Progressive Fish Culturist*, 48:113-9.

Bebee, D., L.E. Dellenbarger, A.R. Schupp, and F. Niami. 1989. "Household Consumption Pattern for Louisiana Crawfish Products," Staff Paper SP-89-10, Louisiana Agricultural Experiment Station, Louisiana State University Agriculture Center, Baton Rouge, Louisiana.

Bell, F.W. 1968. "The Pope and the Price of Fish." *American Economics Review*. 17(5): 1346-50.

Blaylock, J.R., J.M. Feldman, and R.C. Haidache. 1988. "U.S. Demand for Fish: Household Expenditure Demographics and Projections." *Proceedings of the Symposium of Markets for Seafood and Aquacultural Products*, International Institute of Fisheries Economics and Trade and the South Carolina Wildlife and Marine Resources Department. pp. 85-103.

Capps, O. and J. Havlicek. 1984. "National and Regional Household Demand for Meat, Poultry and Seafood: A Complete Systems Approach," *Canadian Journal of Agricultural Economics*. 32:93-107.

Dellenbarger, L.E., J. Dillard, A.R. Schupp, and B.T. Young. 1988. "Socioeconomic Factors Associated with At-Home and Away-From-Home Catfish Consumption in the United States." Unpublished report, Louisiana Agricultural Experiment Station, Baton Rouge, Louisiana. 21 pages.

Ferguson, C.E. and S. Maurice. 1974. *Economic Analysis*. Illinois: Richard D. Irvin, Inc.

Hatch, U. and C. Engle. "Economic Analysis of Aquaculture as a Component of Integrated Agro-Aquaculture Systems: Some Evidence from Panama." *Journal of Aquaculture in the Tropics*. 2:93-105.

Houston, J.E. and A. Nieto. 1988. *Impact of Regional Shrimp Production, Consumer Income, and Imports on Ex-Vessel Prices*. Research Bulletin 377, The Georgia Agricultural Experiment Stations, College of Agriculture, The University of Georgia, Athens, Georgia. p. 25.

Ivers, T.C. 1981. *The Seasonality of Supply and Demand in the Farm-Raised Catfish Industry*, Ph.D. Dissertation, Auburn University, Auburn, Alabama. 116 pages.

Kabir, M. and N.B. Ridler. 1985. "The Demand for Atlantic Salmon in Canada: Reply." *Canadian Journal of Agricultural Economics*. 33:247-9.

Kabir, M. and N.B. Ridler. 1984. "The Demand for Atlantic Salmon in Canada." *Canadian Journal of Agricultural Economics*. (32)560-8.

Kahma, I.H. and G.F. Newkirk. 1987. "Economics of Tray Culture of the Mother-of-Pearl Shell, *Pinctada margaritifera*, in the Red Sea, Sudan." *Journal of the World Aquaculture Society*. 18(3):156-61.

Kingsley, J.B. 1982 "Legal Constraints to Tilapia Culture in the United States." *Journal of the World Aquaculture Society*. 18(3):201-3.

Kinnucan, H. and D. Wineholt. 1989. *Econometric Analysis of Demand and Price-Markup Functions for Catfish at the Processor Level, Alabama*. Bulletin No. 597, Alabama Agricultural Experiment Station, Auburn University, Auburn, Alabama, 30 pages.

Kinnucan, H. 1985. "Demand and Price Relationship for Commercially Processed Catfish with Industry Growth Projections." Paper presented at the Auburn Fisheries and Aquaculture Symposium, September 20-21.

Kinnucan, H., S. Sindelar, D. Wineholt, and U. Hatch. 1988. "Processor Demand and Price-Markup Functions for Catfish: A Desegregated Analysis with Implications for the Off-Flavor Problem." *Southern Journal of Agricultural Economics*, 20(2):81-91.

Kleith, W.R. and F.J. Prochaska. 1988. "Determinants of Household for Seafood and Seafood Quality." *Proceedings of the Symposium of Markets for Seafood and Aquacultural Products*; International Institute of Fisheries Economics and Trade and the South Carolina Wildlife and Marine Resources Department. pp. 104-22.

Lin, B. and N.A. Williams. 1985. "The Demand for Atlantic Salmon in Canada: A Comment." *Canadian Journal of Agricultural Economics*. 33:243-6.

Marshall, Alfred. 1920. *Principles of Economics*, Eighth Ed. Macmillan and Co. Ltd. London, pg. 78.

McCracken, V.A. 1988. "Consumer Demand for Seafood Away-From-Home." *Proceedings of the Symposium of Markets for Seafood and Aquacultural Products*; International Institute of Fisheries Economics and Trade and the South Carolina Wildlife and Marine Resources Department, pp. 124-47.

Pavelec, C.L. 1989. "Fish Consumption and Omega-3 Fatty Acid Intake in Central Alabama Men With and Without a History of Heart Disease." Master of Science Thesis, Auburn University, Auburn, Alabama.

Rupp, E.M., F.L. Miller, and C.F. Baser. 1980. "Some Results of Recent Surveys of Fish and Shellfish Consumption by Age and Region of U.S. Residents." *Health Physics*. 39:165-75.

Shang, Y.C. 1974. "Economic Potential of the Eel Industry in Taiwan." *Aquaculture*. 3:415-23.

Sindelar, S., H. Kinnucan, and U. Hatch. 1987. *Determining the Economic Effects of Off-Flavor in Farm-Raised Catfish*, Bulletin 583, Alabama Agricultural Experiment Station, Auburn University, Auburn, Alabama. 26 pages.

Street, D.R. and G.M. Sullivan. 1985. "Equity Considerations for Fishery Market Technology in Developing Countries: Aquaculture Alternatives." *Journal of the World Mariculture Society*. 16:169-77.

Tomek, G.W. and K.L. Robinson. 1975. *Agricultural Product Prices*, Ithaca: Cornell University Press.

U.S. Department of Agriculture. 1981. *Aquaculture Situation and Outlook*, Washington, D.C.

U.S. Department of Agriculture, Economic Research Service. 1972. *Demand for*

Farm-Raised Channel Catfish in Supermarkets: Analysis of a Selected Market, Marketing Research Report. 993:1-21.

Wineholt, D.A. 1988. *Economic Legal, and Practical Aspects of a Producer Marketing Cooperative for Catfish in West Alabama*, Unpublished Masters Thesis, Auburn University, Auburn, Alabama. 124 pages.

Zidack, W. and U. Hatch. 1991. "An Econometric Estimation of Market Growth for the U.S. Processed Catfish Industry." *Journal of the World Aquaculture Society*, Vol. 22, No. 1, pp. 10-33.

Zidack, W., H. Kinnucan, and U. Hatch. 1989. "A Dynamic Monthly Econometric Model of the U.S. Catfish Industry." Paper presented at the annual meetings of the American Agricultural Economics Association, Baton Rouge, Louisiana. 24 pages.

Chapter 4

Production

PURPOSE OF PRODUCTION

The purpose of production is to satisfy human wants and needs. The basic human wants/needs are for food, clothing, shelter, and security. As civilization modernizes, individual wants and needs multiply. People engage in production to obtain the means by which they will be able to satisfy their own wants, and at the same time help satisfy those of other people. The same people are both producers and consumers. A farmer produces fish, but also purchases other food types from other producers. There is an awareness that as the productivity of one individual increases, that of others also will increase. Thus, the purpose of production is to improve economic welfare, that is, to raise the standard of living and quality of life, by enabling people to satisfy more fully a greater number of their wants. The fish farmer is engaged in production to satisfy human wants for fish, and perhaps recreation. Therefore, before beginning production, farmers must consider wants and needs of others which are expressed in a place of exchange called a market.

Production Defined

Production may be defined as the process of combining resources and forces in the creation of some valuable good or service. Production covers the following activities:

1. changing the form of a good at any stage from the raw material to the finished product, e.g., animal by-products to fish feed;
2. changing the situation of a good, e.g., the production of fish on the farm, and the production of dressed fish at the processing plants;
3. changing the position of the good in time, e.g., over-wintering fish; and
4. the provision of some service, e.g., providing farmers with technical advice through an extension service.

Factors of Production

Production involves several factors which are classed in four main categories. These are called the four factors of production—land, labor, capital, and management.

Land is natural wealth that is used in production. Besides the soil, land also includes native trees, wild animals, minerals, water, streams, and natural ponds. In a broad sense, sunshine and air might be classed as land.

Capital is a produced good used in production, but it differs from land because it is man-made rather than found in nature. All man-made factors such as fertilizer, feed, tanks, sheds, money, and other technologies are considered capital.

Labor is primarily physical energy used in production. Usually it is broadly classified into operator's labor, family labor, and hired labor.

Management refers to the mental energy used in production, as distinguished from physical energy. Management is primarily concerned with decision making and risk bearing.

Fish Production

Fish production is primarily a biological concern. Hence, production involves the combination of aquatic resources, labor, and management to obtain a consumable good. The biologist is concerned with the production response curve for fish as feed, seed stock, water quality parameters, and other factors are varied. The economist is equally concerned with these response curves for establishing the cost efficient level of production. The beginning stage for micro or farm analysis is the biological production function.

THE PRODUCTION FUNCTION

The *production function* is a technical relationship concerning inputs and outputs at a given time using existing technology.

Inputs are productive services, materials and forces used in the production processes. Aquacultural inputs include fingerlings, feed, chemicals, ponds, machinery, and technical, institutional, and organizational services.

Outputs are the goods and services resulting from the processes, which may be thought of as the sum of physical materials and forces. These may include fish, fish-bait, watercress, water chestnuts, and other aquatic products.

The output of fish is somewhat complex and is the result of a wide variety of inputs. The level of output is a function of the level of each of the inputs used as well as any interaction that may occur. A production function for fish may be represented algebraically as:

$$Y_1 = f\,(X_1,X_2,X_3,X_4X_5,\ldots X_n)$$

where:

Y_1 = fish output
X_1 = the amount of feed
X_2 = the stocking rate size of fingerlings
X_3 = the survival rate
X_4 = the stocking density
X_5 = the grow-out period
X_n = some variable related to growth.

This statement merely shows that fish output is related to each of these variables in some way. Other variables may also affect output, but the list includes those usually considered the most important. Restricting the list of variables is to simplify calculations and to measure those variables with significant effects. If these variables are observed for a large number of instances in a controlled experiment, some mathematical formula may be derived which describes the exact relation between the output variable and each input variable. Multiple regression analysis provides one such tool.

Obtaining a set of observations to measure a production function is not always easy. This is particularly so when the detailed impact of one variable, or the interaction between two variables, is sought. The simplest way is merely to experimentally hold all variables except one at a given level. The result might be on the yield response to amount of feed, all other variables held constant. Thus we would have:

$$Y = f(X_1, | X_2, X_3, X_4, X_5, \ldots X_n)$$

Table 4-1 shows the effect of initial stocking density on yield of catfish. Figure 4-1 is a histogram showing the relationship between marketable fish and stocking density in ponds.

Figure 4-1 shows production as a stepped function because of a limited number of stocking densities. Obtaining a continuous curve for production is difficult unless one has many points. Using the given net-production as a function of stocking density and the equation below, a continuous curve may be drawn in contrast to the histogram in Figure 4-1.

TABLE 4-1. Production of Marketable Catfish Using Recommended Practices for Alabama, U.S.A., 1988[a]

Initial stocking pounds/acre	Stocking density fish/acre	Total weight of marketable fish lbs/acre	Change in weight pounds
50	2,500	2,350	--
70	3,500	3,290	940
90	4,500	4,230	940
100	5,000	4,600	370
110	5,500	5,100	876
130	6,500	5,850	744

[a]From 2,500 to 4,500 fish per acre, the farmer experienced a loss of 6.0 percent. As the stocking rate increased, the percent losses were 8.0 for 5,500 per acre and 10.0 for 6,500 per acre.

FIGURE 4-1. Histogram Showing the Net Production of Catfish Under Different Stocking Densities at the End of 200 Days of Production

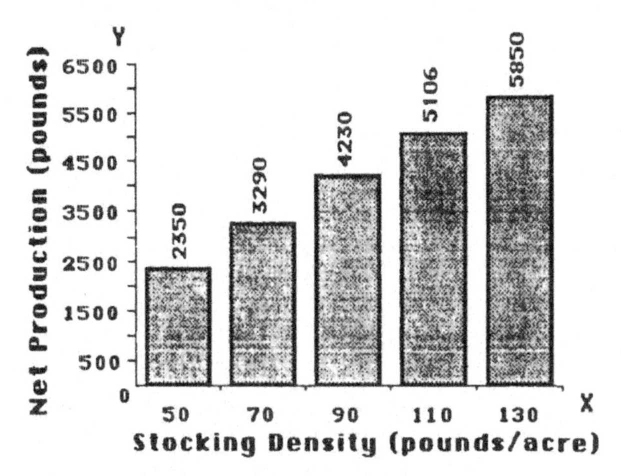

Note: From a stocking density of 50 to 90 pounds per acre, the farmer experienced a loss of 6.0 percent, but the stocking density increased. The percentage losses were 8.0 for 110 and 10.0 for 130.

$$Y_1 = y_1 - a + B_1X_1 - b^2X_1^2$$

Y_1 = net production in lbs.
X_1 = initial stocking density.

The production function is then stated as the relationship between a single product and a single resource describing the output level observed as the level of input is varied. The definition also implicitly specifies that the relation refers to a given production period and a given technology. The more typical curve looks something like that shown in Figure 4-2.

Note that the production function shown in Figure 4-2 has some special characteristics. First, it is observed that output increases with input at an increasing rate. Second, the rate of increase is not constant throughout the curve.

The production function for fish is somewhat complicated. Brett

FIGURE 4-2. Response Curve of Total Marketable Catfish Production (TMP) and Stocking Densities (pounds per acre).

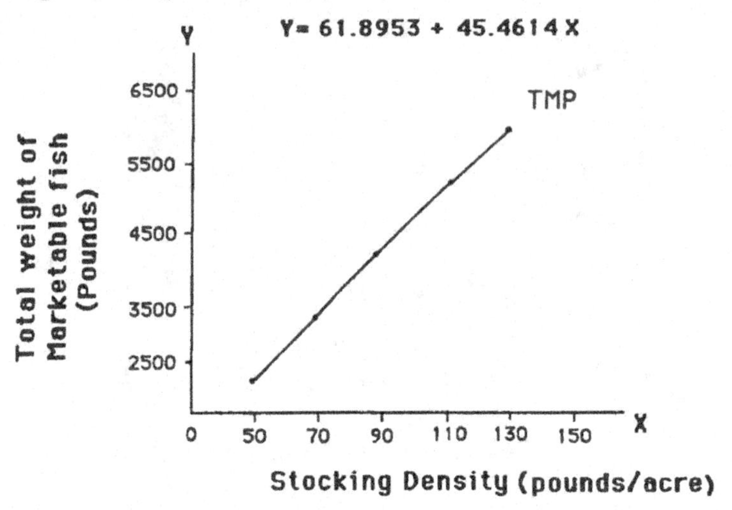

Note: The simple linear equation $Y = 61.8953 + 45.4614X$ represents the best fit for the set of points, $R^2 = 0.99$.

(1979) emphasized the relationship between growth rate and food intake and calls it a "fundamental relation," which holds much of the key to understanding the action of environmental factors (Cacho, 1988). A theoretical feed response curve for catfish was adapted from Brett (see Figure 4-3). This curve shows that the proportion of total energy intake which is available for growth decreases as ration amount increases. Total energy (TE) represents the maximum energy intake by a fish at a given temperature; at this point maximum growth rate is achieved. Maintenance energy (ME) is the point where the growth rate (GR) curve intercepts the horizontal axis; it is the amount of food necessary to maintain fish energy content unchanged under the given environmental conditions. Under optimal water quality conditions, the position of TE and ME will be determined mainly by fish weight and water temperature. One of the more simple production functions observable in the production of certain species, such as catfish, is that of temperature, which has a strong effect on fish growth rates. This fact is well

FIGURE 4-3. Theoretical Feed Response Curve for Catfish

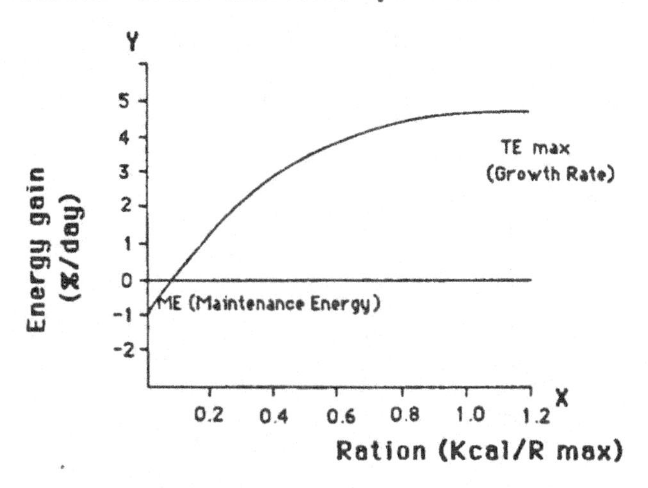

documented for catfish (West, 1966; Stickney and Andrews, 1971; Andrews et al., 1972; Andrews and Stickney, 1972). Beginning at low temperatures, growth rate reaches a maximum at the optimal temperature for the given species, and decreases as temperature increases above the optimal level to reach the lethal level (Figure 4-4).

Thus, production is related to the *law of diminishing returns* which states:

> In the production process, and for all biological processes, when a variable input factor is increased while all other factors remain fixed, production first increases at an increasing rate, then at a decreasing rate, then reaches a maximum and finally declines.

The law of diminishing returns corresponds to the response analysis, the agronomic principle known as "the law of the minimum," formulated by Von Liebig around 1840. This "law" states that "the yield of any crop is governed by any change in the quality of the scarcest factor, called the minimum factor, and as the minimum factor is increased the yield will increase in proportion to the supply of that factor until another becomes the minimum." The law ap-

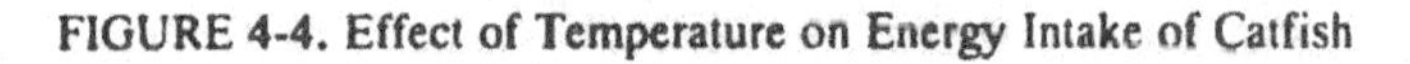

FIGURE 4-4. Effect of Temperature on Energy Intake of Catfish

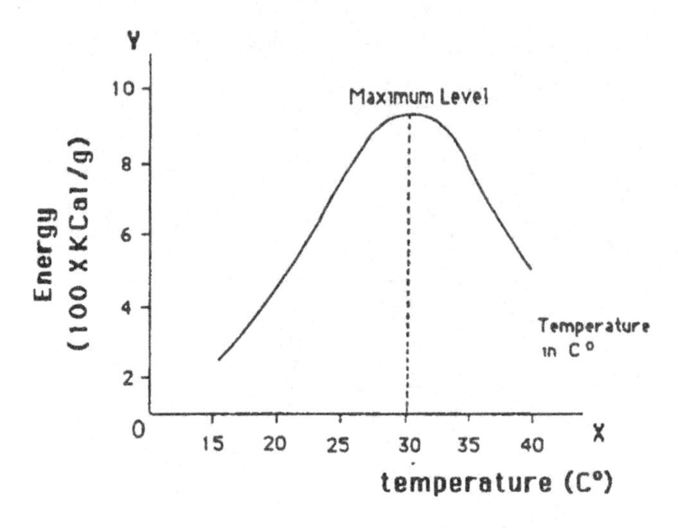

plies to all biological entities. Thus, fishweight, for example, will increase under optimal water quality and temperature conditions only to the extent allowed by the most limiting factor. The most limiting factors may be feed quality, temperature, or other environmental factors.

Hastings and Dupree (1969) developed a response curve for catfish protein level in feed. The response curve increased at an increasing rate, then at a decreasing rate. Observe the demonstration of the law of diminishing returns in Figure 4-5. The response curve for tilapia and several other species to protein level showed the same behavior. As the protein level increases, holding other factors constant, production first increases at an increasing rate, then at a decreasing rate, then reaches a maximum. Reutebuch (1988) found that the growth response curve for catfish when fed four levels of protein could be represented by a quadratic function. He produced two equations, one on protein consumption and live weight gain, and the other on protein consumption and dressed weight gain. Live weight gain and dressed weight gain increased as protein level increased, but both finally declined with higher levels of protein.

FIGURE 4-5. Growth in Aquaria of Channel Catfish Fingerlings Fed Practical Rations Having Different Percentages of Protein

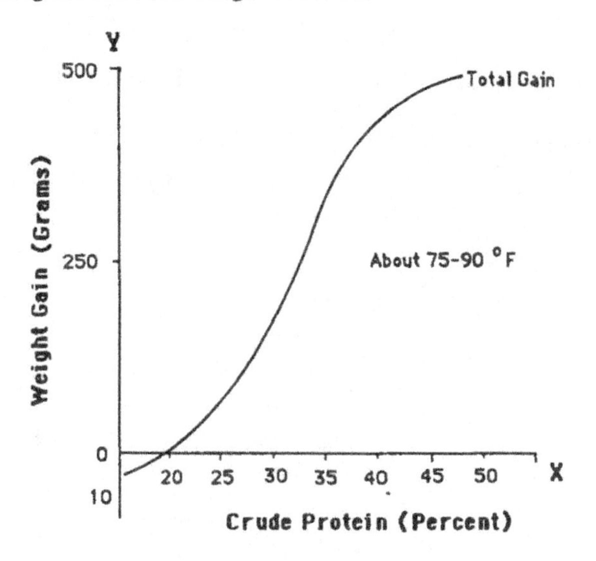

Source: Data for this figure were extracted from a report by W. H. Hastings, and H. K. Dupree, 1969. *Practical Diets for Channel Catfish in U.S.* Interior Bureau of Sport Fisheries, 1968. Resource Publication 77, pp. 224-226.

This research has some limitations in that only four levels of protein were used. However, it gives some indication of the response of fish growth to protein content in feed.

As may be seen from the examples presented over the past few pages, the production function may assume several forms, i.e., linear, quadratic, or hyperbolic. Hence, the production response curve might be represented by a simple or complex equation.

The shape of the response curve of fish is directly observable in the behavior of the marginal product. The term *marginal* means additional or incremental. The marginal product (MP) or marginal physical product (MPP) of factor X_1 is the change in total physical product (TPP) resulting from a one-unit change in X_1. Thus, MPP represents the slope, or the first derivative, of the production function and is calculated:

$$MPP_{x_1} = \frac{\Delta TPP}{\Delta X_1}$$

The MPP is an important production relationship. A typical marginal curve is shown in Figure 4-6.

The average physical product (APP) curve is also shown in Figure 4-5. The APP indicates the amount of output per unit of variable input for various levels of the input. Mathematically this may be expressed as:

$$APP_{x_1} = \frac{TPP}{X_1}$$

FIGURE 4-6. Typical Physical Product, (TPP) Average Physical Product, (APP) and Marginal Physical Product (MPP) Curves

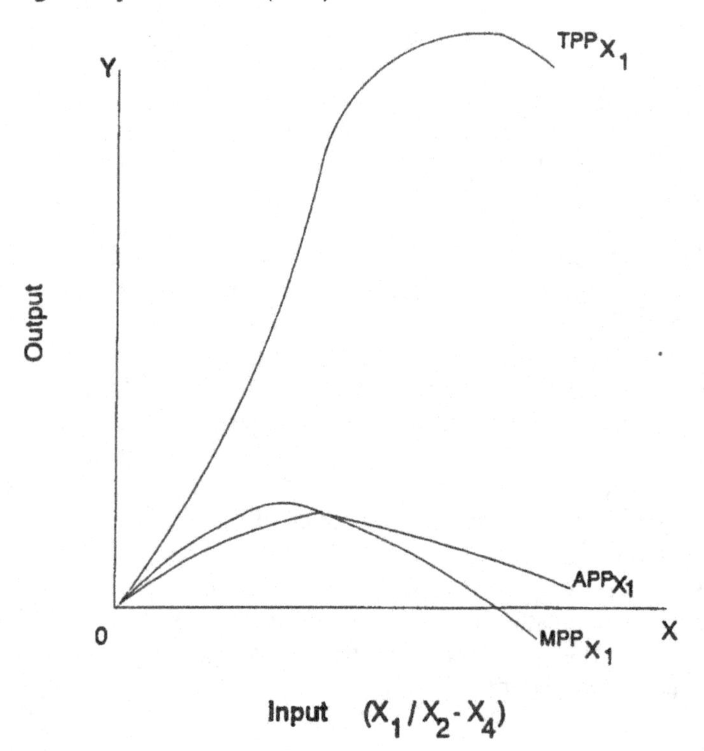

Computation of APP requires dividing total output by the amount of inputs used to obtain that total product. The relationship of APP to MPP is important in analyzing the productive efficiency of the variable input. Efficiency varies with the average physical product. In Table 4-2, the marginal and physical products are calculated for catfish produced in a one-acre pond. Observe that total production increases at an increasing rate then at a decreasing rate and finally declines. Also observe that MPP = APP at stocking densities of 50 and 70 pounds per acre, then APP exceeds MPP.

RELATIONSHIP AMONG TPP, MPP, AND APP

Another look at Figure 4-6 shows that when MPP is greater than APP, average physical product is rising. When MPP is less than APP, average physical product is decreasing. MPP and APP are equal at the point where APP is at a maximum.

When MPP is rising, TPP is increasing at an increasing rate. When MPP is maximum, TPP begins to increase at a decreasing rate. When MPP = 0, TPP is maximum. When MPP < 0, TPP is decreasing. The relationship of the APP and MPP curves is interesting to production economists. These curves are used to subdivide the production function into stages of production.

TABLE 4-2. Total, Marginal, and Average Products for Catfish Stocked at Six Different Densities Per Acre[a]

Stocking density pounds per acre	Stocking density fingerlings per acre	Total weight of fish harvested pounds TPP	MPP	APP
50	2,500	2,350	--	47
70	3,500	3,290	47	47
90	4,500	4,230	47	47
110	5,500	5,106	44	46
130	6,500	5,850	37	45
150	7,500	6,375	26	42

[a]The decrease in the rate of increase of total product is due to increases in the death loss.

Stages of Production

The stages of production are illustrated in Figure 4-7. Three stages of production are then identified.

Stage I is the portion from the origin, 0, to point A, where marginal product crosses the average product. If the farmer knows the

FIGURE 4-7. The Production Function: A Physical Input-Output Relationship and Stages of Production

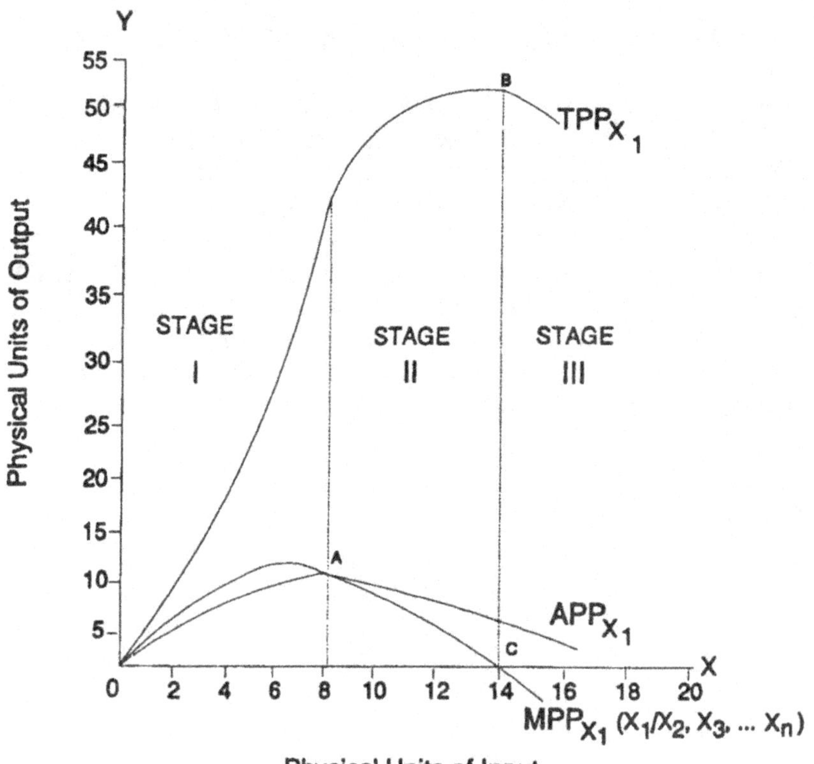

Note: TPP_{X_1} is the Total Physical Product derived from input X_1
APP_{X_1} is the Average Physical Product derived from X_1
MPP_{X_1} is the Marginal Physical Product derived from X_1

probable results and wants to minimize losses, the farmer will not operate within this stage. In Stage I, APP is still increasing. Production should be increased at least to the end of Stage I.

Stage II is the portion from point A to point B, maximum TPP. This stage is the rational area of production. The point of greatest net returns or least loss will occur in this stage. The most profitable point of operation in Stage II cannot be determined unless both resource and product prices are also known. Before going further, note again some observations about the relationships between total product, average product, and marginal product.

1. The average product yield increases as long as the marginal product is above it.
2. The marginal product curve crosses the APP curve at the peak of the APP curve.
3. When the APP is falling, the MPP is below it.
4. The MPP = 0 when TPP is at its peak.

Stage III is the portion to the right of the peak of the total product, or the point where the marginal product is 0. No one wants to operate in Stage III because the total product is less than it could be, but more variable resource is used than when total yield is at a maximum.

Table 4-3 provides an example of the stages of production. Between 6 and 7 units of inputs, MPP and APP are almost equal. Stage I exists between 1 unit and 6 units of labor input. Stage II begins between 6 and 7 units of input and ends somewhere between 10 and 11 units, where MPP approaches 0 and TPP is maximum. Stage III begins between 10 and 11 units of inputs and when MPP = 0.

Decision-Making Rules

Some decision-making pointers may be derived from the production function. For example, it is rational for the farmer to keep increasing input where total physical product is increasing at an increasing rate, or where average physical product increases by increasing input (Stage I of the production function). Also, it is not

TABLE 4-3. The Stages of Production

Stages	Units of labor	Total product	Average product	Marginal product
	1	10	10	10
	2	24	12	14
Stage I	3	40	13	16
	4	56	14	19
	5	75	15	19
	6	90	15	15
	7	103	14.7	13
	8	112	14.0	9
Stage II	9	119	13.2	7
	10	120	12.0	1
Stage III	11	118	10.6	-2

profitable to increase inputs that would decrease total physical product (Stage III of the production function). If TPP decreases, marginal physical product becomes negative. Consequently, there is only one rational area under the production function for decision making. That area is between maximum APP and maximum TPP, or between equality of MPP and APP and zero MPP (Stage II of the production function). A farm should always strive to produce in Stage II.

OPTIMIZING THE USE OF A SINGLE RESOURCE

Optimization in a limited sense is attainment of that input level which maximizes net income from the use of the resource. Net income is the difference between total revenue (TR) and total cost (TC). Mathematically, profits will be optimized when value of the marginal product (VMP) is equal to the price of the input (P_x). The value of the marginal product is:

$$VMP = MPP \cdot P_y$$

Therefore, profit optimization is reached when:

$$\begin{aligned} MPP \cdot P_y &= P_x \text{ or} \\ MPP &= P_x/P_y \end{aligned}$$

where:

P_y = price of the output
Px = price of the input.

Certain assumptions must be made when optimizing output with a single resource. Assume for the moment that inputs are unlimited and that the purchase of inputs and sale of output are undertaken in a perfectly competitive market situation. Assume also that we are dealing with a small production system with a 0.1 hectare (0.25 acre) pond where fish (in kilograms) is the only output.

In Table 4-4 the single variable input is feed (bags of 20 kilograms [44 pounds] each). All other inputs (land, labor, stocking rate, etc.) are assumed to be fixed, bags of feed are available in unlimited quantity, and the producer has no capital constraint. Feed is assumed to have a constant cost (P_x) of $8.00 per bag and the farm gate price (P_y) for fish is $2.00 per kilogram (2.2 pounds). Assume the output price does not change in response to increases in

TABLE 4-4. Hypothetical Data Showing Profit Maximizing Principle When Inputs Are Unlimited

Bags of feed	TPP	APP	MPP	VMP	Marginal cost (P_x)	TR (TVP)	TC	Profit
0	0	0	0	0	0	0	0	0
1	5	5	5	10	8	10	8	2
2	15	7.5	10	20	8	30	16	14
3	26	8.7	(11)	22	8	52	24	28
4	35	(9.0)	9	18	8	70	32	38
5	41	8.2	6	12	8	82	40	(42)
6	44	7.3	3	6	8	88	48	40
7	46	6.6	1	2	8	(92)	56	36
8	45	5.6	-1	-2	8	90	64	26
9	43	4.8	-2	-4	8	86	72	14

output from our small producer. The small producer is a price taker operating in a competitive market. The question the fish farmer is trying to answer is "how many bags of feed may I use to maximize profits?"

Maximum profits ($42) are earned when five bags of feed are used. At lower levels of input use, the value of the marginal physical product (VMP) obtained from each added input is greater than the marginal cost P_x of the added input. Beyond five bags of feed, the marginal cost exceeds the value of the marginal product of a bag of feed. In other words, the producer should keep adding inputs as long as the additional revenue obtained exceeds additional cost.

There are several interrelated conclusions from this unconstrained case:

1. Maximizing production does not maximize profits. For example, maximum production is achieved with seven bags of feed, but profits are lower, $36, than the $42 obtained from using only five bags of feed. Maximum profits are, therefore, obtained at lower levels of output and input use than those that maximize production.
2. The profit maximizing rule is based on marginal principles. A producer who bases production decisions upon average or total production and revenue principles will earn less profit than a producer who uses marginal analysis.
3. The level of fixed cost does not influence the decision of the producer regarding optimal use of the variable input. NOTE that the producer's decision is based upon a comparison of the *value* of Marginal Product and Marginal Input.

EMPIRICAL PRODUCTION FUNCTIONS

Aquacultural production is a rather complex biological process. Experimentally, it has been observed that a functional relationship exists between certain inputs and outputs, which gives rise to a production function with all three stages of production. However, biological laws of growth are not uniform. Therefore, a mathematical function which best explains the relationship between inputs and output may be a poor device for explaining the input-output phenomenon. Production functions may be obtained using empirical data only with considerable effort and even at best, the equation obtained is only a mere approximation. Nerrie (1987) developed several production functions for catfish farmers in Alabama. One equation obtained was:

$$Y = .45 + 0.36X_1 + 4.46X_{2a} - 9.27X_{3a} + 0.28X_{4a} - 0.74X_5$$

where:

Y = net production in lbs. per acre per day
X_1 = feed in lbs. per acre per day
X_{2a} = annual capital recovery charge in dollars per acre per day
X_{3a} = production labor in hours per acre per day
X_{4a} = fingerlings stocked per acre per day
X_5 = crop length in days.

This function measures the effects of a few variables on output. For example, feed in pounds per day per acre are shown to positively influence production. Annual capital recovery charge, plus fingerlings stocked per acre positively affect production, while crop length in days and production labor negatively affect production.

Van Dam (1990) derived growth models for the Nile tilapia, *Oreochromis niloticus*, from fish-rice experimental data. Production in terms of gross fish yield (fish biomass stocked in kilograms per hectare), net fish yield (fish biomass harvested minus fish biomass stocked in kilograms per hectare), and fish growth rate (in grams per day) was expressed as a function of 31 independent variables. These variables included period of growth, or crop length in days stocking density, stocking size in grams, basal nitrogen application in kilograms per hectare, n/basal phosphorus application in kilograms per hectare, number of insecticide trials, and the maximum air temperature.

The model shows that period of growth, stocking density, stocking size, nitrogen application, and temperature positively affected gross fish yield; phosphorus had a negative effect on gross fish yield.

Elasticity of Production

The elasticity of production measures the relative change in output in response to a change in input. The elasticity of production, like any other elasticity, is independent of units of measure. Elasticity of production (E_p) is defined as:

$$E_p = \frac{\text{Percentage change in output}}{\text{Percentage change in input}}$$

The elasticity of production is then determined to be:

$$E_p = \frac{\frac{\Delta Y}{Y}}{\frac{\Delta X}{X}} = \frac{\Delta Y}{\Delta X} * \frac{X}{Y} = \frac{MPP}{APP}$$

From Figure 4-1 in Stage I, MPP > APP, therefore, $E_p > 1$. In Stage II, MPP < APP and $E_p < 1$ but > 0. In Stage III, MPP is negative and E_p is negative.

Input Indivisibility

Inputs in aquacultural production are usually available in whole units and are considered indivisible. For instance, neither an aerator nor a tractor can be divided. If one is using one aerator for a one-acre pond, the aerator cannot be divided when the size of the pond is reduced to one-half acre.

In some cases, inputs like aerators or tractors are available in several different sizes. A small model aerator or tractor cannot do the job that a larger model is able to do. Similarly, buildings cannot be divided, though building services may be obtained in different amounts. However, when a farmer purchases the input, some specific model or size, etc., is acquired. That unit then becomes indivisible to the farmer.

Labor in many countries may be employed only in lump sums. Strict employment codes in many developed countries do not allow an employer to hire labor by the hour. Once a person is employed, payment for a full day must be made. The result is labor use in discrete (lumpy) units since the individual may be employed for a day or not at all.

Lumpiness in the use of inputs further complicates the determination of an empirical production function. Curves drawn from production functions involving lumpy inputs are linear, with jumps in the form of steps, as shown in Figure 4-8. Thus, when optimal input/output levels are estimated, the result will be, say, hire 5 laborers or use 1 aerator; whereas a continuous function would call for 4.2 laborers and 1.5 aerator, which is impossible.

FIGURE 4-8. Total Physical Product Derived from an Indivisible Input X

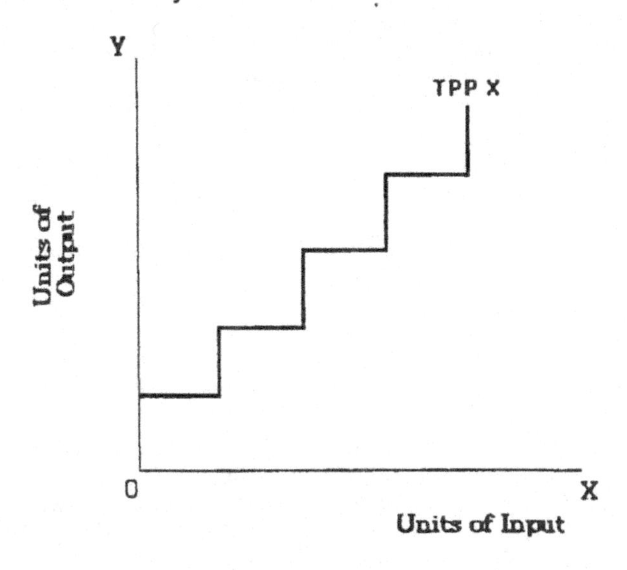

REFERENCES AND RECOMMENDED READINGS

Albrecht, W.P., Jr. 1983. *Economics*. Fourth Ed. Englewood Cliffs, NJ: Prentice-Hall.

Andrews, J.W. and R.R. Stickney. 1972. "Interaction of Feeding Rates and Environmental Temperature on Growth, Food Conversion and Body Composition of Channel Catfish." *Transactions of the American Fisheries Society*. 101:94-9.

Andrews, J.W., L.H. Knoght, and T. Musia. 1972. "Temperature Requirements for High Density Rearing of Catfish From Fingerlings to Market Size." *Progressive Fish Culturalist*. 34:240-1.

Brett, J.R. 1979. "Environmental Factors and Growth," in *Fish Physiology*, vol. 8, W.S. Hoars, D.J. Randall, and J.R. Brett, eds. pp. 599-675.

Cacho, O.J. 1988. "A Bioeconomic Model for Catfish Production with Emphasis on Nutritional Aspects." An unpublished Ph.D. Dissertation, Auburn University, Auburn, Alabama. 149 pages.

Chong, K.C. and M.S. Lizarondo. 1981. "Input-Output Relationships of Philippine Milkfish Aquaculture." Proceedings of a workshop held in Singapore, International Development Center, Ottawa, Ontario, Canada.

Dehadrai, P.V. and P.K. Mukhopadhyay. 1987. "Effects of Urea Supplementation in the Feed of Walking Catfish (*Clarias batrachus*)." *Journal of Aquaculture in the Tropics*. (2):1-7.

Dellenbarger, L.E. and E.J. Luzar. 1988. "The Economics Associated with Crawfish Production from Louisiana Atchafalaya Basin." *Journal of the World Aquaculture Society*. 19,2:41-6.

Dupree, H.K. and J.E. Halver. 1970. "Amino Acids Essential for the Growth of Channel Catfish (*Ictalurus punctatus*)." *Trans. American Fish Society*. 1:90-2.

Greenland, D.C., H.K. Dupree, and C.L. Long. 1984. "Growth Studies Conducted on Channel Catfish Fed Diets with Fish Meal or Fish Meal Substitute." *Feedstuff*. 56(46):20.

Hastings, W.H. and H.K. Dupree, 1969. *Practical Diets for Channel Catfish in U.S.* Interior Bureau of Sport Fisheries and Wildlife Progress in Sport Fisheries Research, 1968. Resource Publication 77, pp. 224-6.

Karp, L., A. Sadeh, and W.L. Griffin. 1986. "Cycles in Agricultural Production: The Case of Aquaculture." *American Journal of Agricultural Economics*. August, 553-61.

Lipschultz, F. and G.E. Krantz. 1980. "Production Optimization and Economic Analysis of an Oyster (*Crassostrea Virginia*) Hatchery on the Chesapeake Bay, Maryland, U.S.A." *Proceedings of the World Mariculture Society*. 11:580-91.

Morita, S.K. 1977. "An Econometric Model of Prawn Pond Production." *Journal of the World Mariculture Society*. pp. 741-5.

Nerrie, Brian. 1987. *Input-Output Analysis of West-Central Alabama Catfish Production*. Unpublished Ph.D. Dissertation, Auburn University, Alabama.

Panayotou, T., S. Wattanutchariya, S. Tsvilanonda, and R. Tokrisna. 1982. "The Economics of Catfish Farming in Central Thailand." *ICLARM Technical Reports 4*, Manila, Philippines.

Rauch, H.E., Botsford L.W., and R.A. Shleser. 1975. "Economic Optimization of an Aquaculture Facility." *IEEE Transactions on Automatic Control*. AC-20(3):310-319.

Reutebuch, E.M. 1988. "Quadratic Growth Response of Pond-Raised Channel Catfish to Increase Levels of Dietary Protein and its Economic Implications." Unpublished Masters Thesis, Auburn University, Auburn, Alabama. 50 pages.

Shang, Y.C. and T. Fujimura. 1977. "The Production Economics of Freshwater Prawn (*Macrobrachium rosenbergii*) Farming in Hawaii." *Aquaculture*. 11:99-110.

Smith, I.R. 1982. "Microeconomics of Existing Aquaculture Production Systems: Basic Concepts and Definitions," in *Aquaculture Economics Research in Asia*. Proceedings of a workshop held in Singapore, 2-5, June 1982.

Smith, T.I.J., P.A. Sandifer, W.E. Jenkins, and A.D. Stokes. 1981. "Effect of Population Structure and Density at Stocking on Production and Commercial Feasibility of Prawn (*Macrobrachium rosenbergii*) Farming in Temperate Climates." *Journal of the World Mariculture Society*. 12(1):233-50.

Spivey, W.A. 1973. "Optimization in Complex Management Systems." *Transactions of the American Fisheries Society*. 2:492-9.

Stickney, R.R. and J.W. Andrews. 1971. "Combined Effects of Dietary Lipids and Environmental Temperature on Growth, Metabolism and Body Composition of Channel Catfish (*Ictalurus punctatus*)." *Journal of Nutrition*. 101:1703-10.

Stickney, R.R. and R.T. Lovell. 1977. *Nutrition and Feeding of Channel Catfish*, A Report from the Nutrition Subcommittee of Regional Research Project S-83, Bulletin 218, pp. 67.

Talpaz, H. and Y. Tsur. 1982. "Optimizing Aquaculture Management of a Single-Species Fish Population." *Agriculture Systems*. 9:127-42.

Thia-Eng, C. and T. Seng Keh. 1978. "Economic Production of Estuary Grouper, (*Epinephelus salmoides*) Maxwell, Reared in Floating Net-Cages." IFS Regional Meeting on Aquaculture, Malaysia: Penang. 104-86.

U.S. Department of the Interior Fish and Wildlife Service. 1984. *Third Report to the Fish Farmers: The Status of Warmwater Fish Farming and Progress in Fish Farming Research*. Edited by H.K. Dupree and J.V. Huner; U.S. Fish and Wildlife Service, Washington, D.C.

Van Dam, A.A. 1990. "Multiple Regression Analysis of Accumulated Data From Aquaculture Experiments: Rice-Fish Culture Example." *Aquaculture and Fisheries Management*. 21:1-15.

West, B.W. 1966. "Growth, Food Conversion, Food Consumption and Survival at Various Temperatures of the Channel Catfish (*Ictalurus punctatus*) (Rafinesgue)." Unpublished M.S. Thesis, University of Arkansas, Fayetteville, Ark. 65 pages.

Willis, S.A. and M.E. Berrigan. "Effects of Stocking Size and Density on Growth and Survival of (*Macrobrachium rosenbergii*) (De Man) in Ponds." *Proceedings of the World Mariculture*. 8:251-64.

Chapter 5

Cost of Production

When the fish farmer decides to engage in production, the resource requirements and the price of those resources must be determined. Resources have many uses, but when a given set of resources, say land and labor, is used to produce fish, it cannot be used at the same time in the production of corn. If the resources used to produce fish had only one alternative use—the production of corn—the value of resources used to produce a specific amount of fish would be the value of those same resources in corn production. If, however, the resources required to produce the fish could be used to produce many different things, their value is derived from the most highly valued product that cannot be produced, because the resources are used for fish production. In other words, the cost of anything is the value of the best alternative given up in order to get that thing. The sacrificed alternative is referred to as the *opportunity cost*.

The alternative or opportunity cost of producing one pound of fish is also called the *social cost* of production. If there were a market for the value of resources used in the production of a good, the social costs would equal the sum of the payments for the resources used. The total costs of production, therefore, consist of those payments necessary to attract and keep the factors of production attached to the farm or firm, including the human resources of the entrepreneur.

The total cost of production is often divided into explicit and implicit costs. *Explicit costs* are accounting expenses. The money payments for fertilizer, fingerlings, chemicals, feed, and other inputs are explicit costs. Explicit costs also include payments for fixed assets, depreciation, and losses incurred in production. *Implicit costs* are opportunity costs that are not reflected in the farmer's accounting statement. These costs must be considered in developing countries where labor use is intensive. Fish production involves fewer market inputs, but much family labor. The implicit costs of family labor in other jobs must be accounted for if one must determine how costs affect production decisions. Another type of implicit cost is the use of the farmer's own resources, such as land and capital, in the production process. If returns are not made to land, labor, and family labor profits to the fish enterprise might be overestimated.

SHORT-RUN PRODUCTION COSTS

In general, the cost of producing fish is the sum of the payments made to acquire resources. Costs are derived by applying input prices to the factors of production.

Total Cost

Total cost (TC) is the amount of money that must be expended to obtain various levels of production. The cost of producing the same species of fish varies from region to region because of the differences in climatic and topographic conditions, in the technology used, the distance of the farm from markets of inputs and output, and variation in prices over geographic areas. Cost also varies from farm to farm based on management skill, farm size, and technology. Cost is divided into two groups, *fixed cost* (FC) and *variable cost* (VC).

Fixed Costs

Fixed costs (FC) are those that must be paid regardless of whether the farmer engages in production. These costs include land, property taxes, depreciation, and interests on capital investment, such as drainage and pond construction. Fixed costs do not change in magnitude as the amount of output of the production process changes.

Variable Costs

Variable costs (VC) include payments for items used in production. Variable costs vary during the production period. Total variable costs (TVC) are computed by multiplying the amount of variable input used by the price per unit of input. Total variable costs include payments for items such as feed, fingerlings, fuel, chemicals, labor, and interest on variable payments. The distinction between fixed and variable cost is not always clear. For example, electricity cost may be divided into portions, one of which the farmer must pay even if there is zero production. That is the cost of installation and simply having electricity on the premises. The other portion is electricity used to operate equipment in producing fish.

Costs are related to the quantity of output produced. The relationship between costs and output is shown in Figure 5-1. The total fixed cost (TFC) curve is parallel to the horizontal axis. This means that cost is the same for the production period. The total variable

FIGURE 5-1. Short-Run Total Fixed Cost, Total Variable Cost, and Total Cost Curves

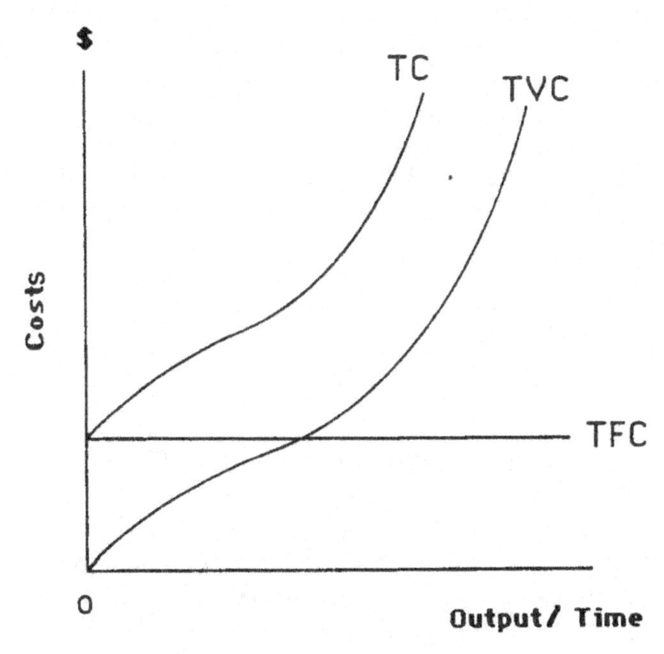

cost (TVC) curve begins from the origin and increases at an increasing rate and then at a decreasing rate. The total cost (TC) is the sum of the TVC and TFC. Thus, the TC curve originates at the point where the TFC curve meets the vertical axis then runs parallel to TVC, because fixed costs do not vary in the short-run; that is, TFC is linear. The TVC curve is the inverse of the production curve discussed in Chapter 4.

Marginal Cost

The relationship between the cost functions and output may be better studied by examination of marginal cost curves. *Marginal cost* (MC) is the additional cost necessary to produce one more unit of output. Marginal costs depend on the nature of the production function and the unit costs of the variable inputs. The shape of the marginal cost curve, shown in Figure 5-2, is derived from the total variable cost curve.

The theoretical MC curve is U-shaped. As additional units of output are produced, MC falls, reaches a minimum, and then rises. MC is high initially because the fixed plant and equipment are not designed to produce very low rates of output and production is more expensive when output is low. MC falls because production efficiency increases as output increases, up to a point, then rises because the plant becomes overutilized as output is increased. Inefficiency in production occurs. The shape of the MC curve is attributed to the law of diminishing marginal returns. MC is calculated as:

$$MC = \frac{\Delta TVC}{\Delta Q}$$

Note that MC is not expressed in terms of TC. TC = TVC + TFC. Since TFC does not vary in the short-run, the change in TFC = 0. Therefore, MC is only a function of TVC. However, under long-run conditions, fixed cost may become variable. Under those conditions, MC is a function of TC. This point is discussed further in the following paragraphs.

FIGURE 5-2. Short-Run Average Variable, Average Total, and Marginal Cost Curves

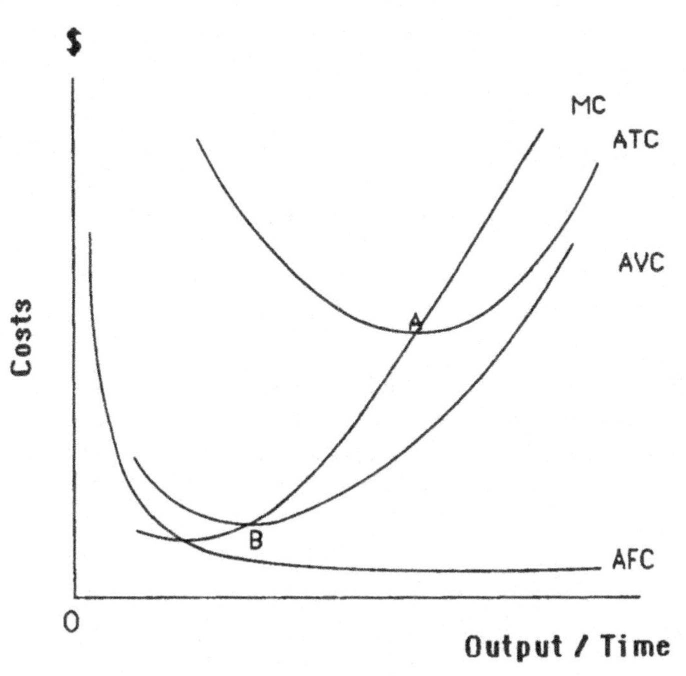

Average Costs

There are three average cost curves: the average total cost (ATC), the average variable cost (AVC), and average fixed cost (AFC) curves (see Figure 5-2).

Average Total Cost

Average total cost is the sum of AVC and AFC:

$$ATC = AVC + AFC$$

ATC is also calculated by dividing TC by output:

$$ATC = \frac{TC}{Q}$$

The ATC curve first declines, reaches a minimum where it intersects the MC curve at point A, and finally increases (Figure 5-2). The ATC measures production efficiency in costs per unit of output. The most efficient level of production is that which produces the lowest cost per unit of output.

Average Variable Cost

Average variable cost is TVC divided by output:

$$AVC = \frac{TVC}{Q}$$

The AVC curve is U-shaped, as dictated by the law of diminishing marginal returns. The AVC curve first declines, reaches a minimum, where it intersects the MC curve at point B, and then rises (Figure 5-2). Note that the minimum of the AVC curve is reached before the minimum of the ATC curve.

Average Fixed Cost

Average fixed cost is TFC divided by output:

$$AFC = \frac{TFC}{Q}$$

The AFC curve declines continuously, approaching the X axis asymptotically (Figure 5-2). AFC may be described as a rectangular hyperbola.

Marginal-Average Cost Curve Relationship

There are some existing relationships between the MC curve and the AVC and ATC curves which are similar to the MPP and APP curves.

1. When the MC curve is below the AC, AC will decline.
2. When MC is above average cost, average cost rises. Therefore, when average cost is rising, marginal cost must be above average cost.

3. When average cost is at a minimum, marginal cost is equal to average cost.

Observe the relationships in Table 5-1. Notice that after MC becomes equal to both AVC and ATC, each begins to increase.

SIMULATED COST CURVES FOR CATFISH

The shape of the TVC and TC curves are not always curvilinear. In many real life situations, the curves could assume a variety of shapes, even straight lines. In the theoretical example given in Table 5-1, it was assumed that in the short-run, output could be increased within a given size plant. All other production factors were designed for the given set of plant and equipment.

In the example, it was further assumed that as production increased, losses would result from problems in management, and from fish mortality caused by oxygen problems. Low levels of dissolved oxygen cause decline in both weight gain and feed conversion. Also, prolonged exposure to low dissolved oxygen stresses fish and makes them more susceptible to bacterial infections (Engle, 1988). This in turn lowers the production efficiency. Therefore, the cost functions would assume curvilinear relationships. The values for the production system are given in Table 5-2. The system

TABLE 5-1. Hypothetical Short-Run Costs for a Catfish Producer

Output (1)	TFC (2)	TVC (3)	TC (4)=(2)+(3)	MC (5)	AFC (6)	AVC (7)=(3)÷(1)	ATC (8)=(4)÷(1)
lbs.	-------	-------	-------	Dol. per acre	-------	-------	-------
0	500	0	500				
1000	500	900	1400	.90	.50	.90	1.40
2000	500	1300	1800	.40	.25	.65	.90
3000	500	1650	2150	.35	.16	.55	.71
4000	500	2000	2500	.35	.12	.50	.62
5000	500	2150	2650	.15	.10	.43	.53
6000	500	2640	3140	.49	.08	.44	.52
7000	500	4200	4700	1.56	.07	.60	.67
8000	500	6000	6500	1.80	.06	.75	.81
9000	500	8550	9050	2.55	.06	.95	1.01
10000	500	12500	13000	3.95	.05	1.25	1.30

TABLE 5-2. Estimated Production Systems, Stocking Rate, and Simulated Costs and Returns for Catfish, 1990

System size	Stocking rate	Total fixed cost	Total variable cost	Total cost	Net returns
Acres	No.	----------	----------	Dol. per acre ----------	----------
1.0	3500	943.55	1363.12	2306.67	-168.17
2.5	3500	1167.98	3237.20	4405.18	376.43
5.0	3500	1235.49	6236.69	7572.18	624.06
7.5	3500	2549.93	9479.74	12029.68	534.54
10.0	3500	2696.60	12515.81	15212.42	617.26
12.5	3500	2843.27	15551.89	18395.16	666.89
15.0	3500	2989.94	18587.96	21577.90	669.97
20.5	3500	4156.77	24823.23	28980.00	689.50

System size	Stocking rate	Average fixed cost	Average variable cost	Average total cost
1.0	3500	28.68	41.43	70.11
2.5	3500	14.20	39.36	53.56
5.0	3500	8.12	37.91	46.03
7.5	3500	10.33	38.43	48.75
10.0	3500	8.20	38.04	46.24
12.5	3500	6.91	37.82	44.73
15.0	3500	6.06	37.67	43.72
17.5	3500	6.97	37.86	44.82
20.0	3500	6.32	37.73	44.04

size is shown to vary from 1.0 acre to 20.5 acres while the stocking rate remains constant. In the simulated model, the same aerator size was assumed for the 1.0-acre to the 7.0-acre size pond and the pump size was assumed to double. The shapes of the resulting TC, TVC, and TFC curves are shown in Figure 5-3. The TFC curve is horizontal and parallel to the X-axis. The TC and TVC curves are slightly curvilinear.

The AFC, AVC, and MC curves for a catfish enterprise are shown in Figure 5-4. Notice the AVC and ATC curves first decline and then intersect the curves at their minimum points.

Profit Maximization

The MC and marginal revenue (MR) curves are the most important relationships to consider when deciding on the quantity of output a farmer or firm should produce to maximize profit. Marginal cost is the additional cost associated with each additional unit of output, while marginal revenue reflects the additional return gener-

FIGURE 5-3. Theoretical Total Costs Functions for Catfish Production

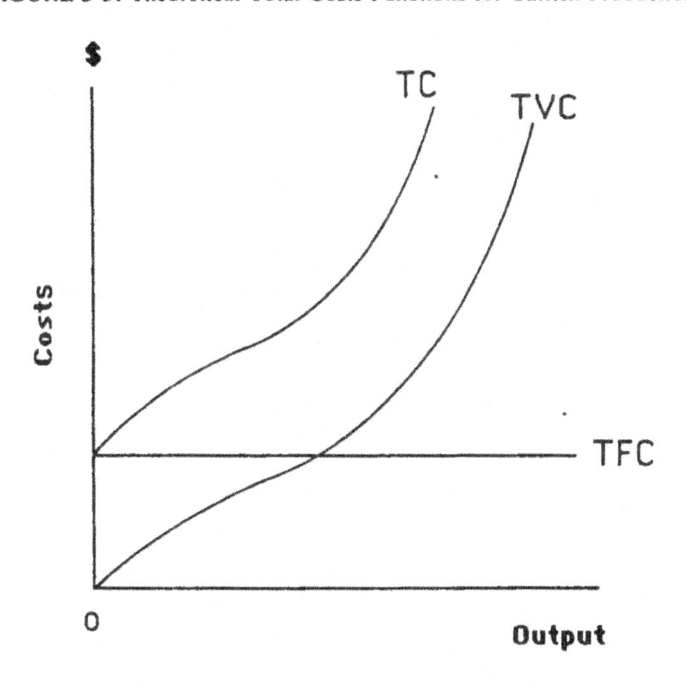

ated for each unit of product. The change in total revenue (TR) associated with a one-unit change in output is marginal revenue. The level of production at which these values are equal (MR = MC) is the profit maximizing output. In Figure 5-5, a perfectly competitive firm is assumed, in which MR is equal to the price. (The concept of a perfectly competitive firm will be studied more thoroughly in a following chapter.) Under this condition, the producing firm is called a price taker. There is no choice but to sell a product at the prevailing market price. Therefore, the MR curve is a straight line parallel to the horizontal axis. At the point where MR = MC, quantity Q* will be produced, and profit will be maximized. Total revenue may be determined by multiplying Q* by p*.

$$
\begin{aligned}
TR &= Q^* \times p^* = OP^*AQ^* \\
TC &= Q^* \times p' = OP'BQ^* \\
\text{profit} &= TR - C = ABP'P^* \\
TVC &= Q^* \times p' = OP'CQ^*
\end{aligned}
$$

FIGURE 5-4. Simulated Average and Marginal Cost Curves for Catfish Production in Alabama, 1988

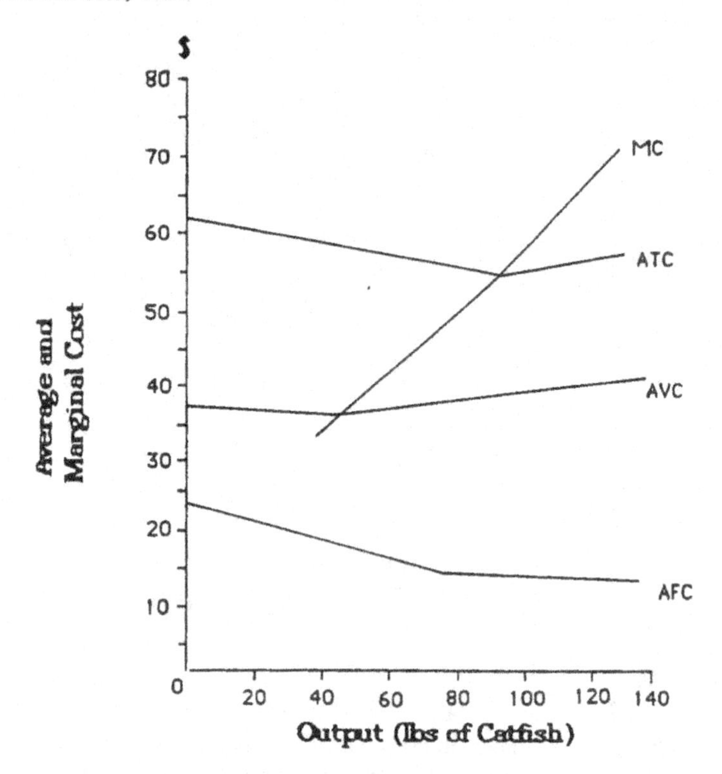

The relationship between marginal cost and marginal revenue may be used to derive the short-run supply for a firm. Supply is defined as the quantities of a good that producers are willing to sell at varying prices at a given point in time, with all other factors held constant. In Figure 5-6, as P is increased from P_1 to P_4, the quantity of output supplied increases from q_1 to q_4. In the short-run, the rational farmer will produce as long as the AVC is covered, but will not produce if the MR is less than AVC. The farmer will produce in the short-run if MR is greater than the minimum variable cost, hoping to cover all costs in a longer time period. As P increases above the minimum average variable cost, quantity supplied increases each time price increases. Thus, the portion of the MC curve above the AVC curve is the short-run supply curve for the farmer.

FIGURE 5-5. Per Unit Cost Curves and Marginal Revenue Curve Used in Factor Product Model

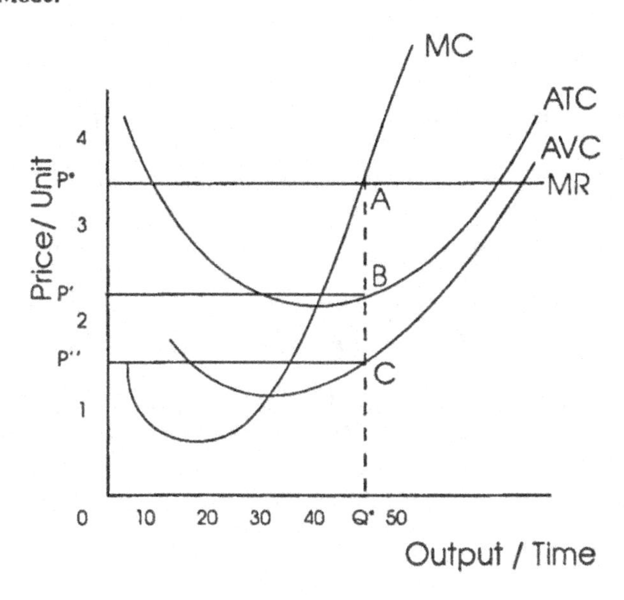

LONG-RUN COSTS

Planning Horizon

The entrepreneur who is beginning a firm for fish processing is making long-run plans. The long-run means that all inputs are variable to the entrepreneur. There is time available to change any cost conditions which the firm may face. The first step in planning a firm is to decide on the plant size. Since the fish farmer has sufficient time to make adjustments to the operation, new facilities may be constructed, new equipment purchased, or existing buildings repaired. The fish processor, on the other hand, must determine the minimum cost of processing a given quantity of fish before deciding on the size or scale of plant. The period of time accepted for planning, developing, and operating a given plant size in the foreseeable future is the planning horizon. The *planning horizon* is the long-run in which nothing is fixed to the entrepreneur. Given factor prices and technology production may be assumed for the period considered by the firm operator. But, these items also may change in the long-run.

The Long-Run Cost Curve

Long-run cost is the cost at which each quantity of output may be produced when no resource is fixed in quantity or rate of usage. The shape of the long-run total cost curves is similar to that of the short-run cost curves. The slope of the long-run total cost (LRTC) curve depends upon the production level and prevailing factor prices (see Figure 5-7).

Long-Run Average Cost and Long-Run Marginal Costs

Long-run average cost is the total cost of producing a particular quantity of output divided by the quantity. Since all long-run costs are variable, the average variable cost is very important in determining plant size. In the short-run, cost curves were assumed opti-

FIGURE 5-6. Illustration of Supply Curve Derived from Marginal Cost

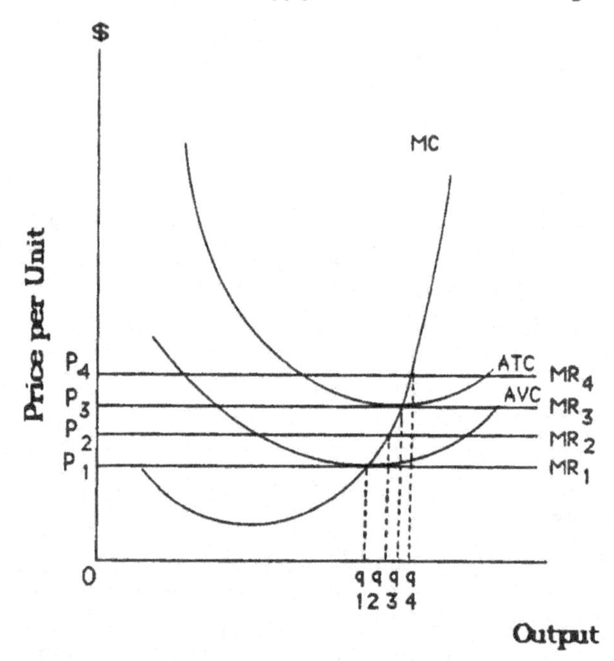

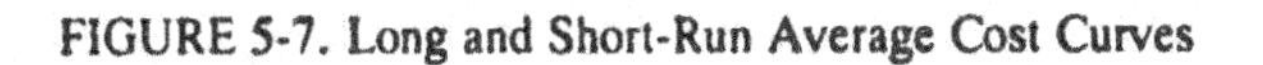

FIGURE 5-7. Long and Short-Run Average Cost Curves

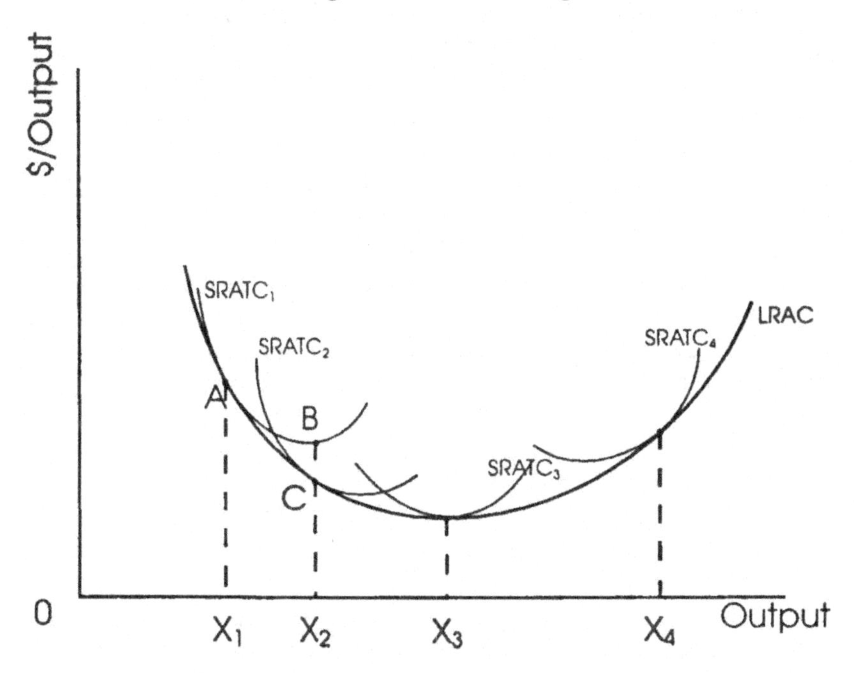

mal. Thus, the costs of input combinations for producing any level of output, given the existing plant and technology, were considered the *least* costs.

The long-run average cost (LRAC) curves may be considered an envelope of the short-run average cost (SRAC) curves for optimal plants of various scale. In Figure 5-7, short-run average cost curves representing different scales of production are shown. Plant size A provides the least cost for producing $0X_1$, plant size B provides the least cost for producing quantity $0X_2$ and so on. The tangent of each SRAC curve with LRAC indicates the minimum long-run average cost of producing each level of output. When there is an innumerable number of plants a U-shaped long-run average cost curve is formed.

Technological factors also contribute to economies of scale. The productivity of equipment usually increases with size.

On the market side, the availability of discounts leads to economies through large-scale purchasing of raw materials, supplies and other inputs. These economies extend to the cost of capital because large firms typically have greater access to capital markets and can acquire funds at lower rates. These factors and others lead to increasing returns to scale and thus decreasing average costs.

It is important to observe that the U-shaped long-run average cost curve is the least cost curve of all firms combined. But, each level of output typically will not be produced where its average short-run average cost curve is minimized. Plant A's short-run average cost curve is minimized at point M, but at that output Plant B is more efficient. That is, Plant B's short-run average costs are lower. Where increasing returns to scale exists, the least cost plant usually operates at less than full capacity. Capacity is the point where short-run average costs are minimized and is the scale of plant which produces an output rate that provides the entrepreneur with no incentive to change the plant scale.

Long-run marginal costs reflect all costs in the long-run since there are no fixed expenses. As in the short-run, a long-run marginal cost curve (LRMC) will be U-shaped and intersect the LRAC from below at its minimum point, shown in Figure 5-8. Note that with the longer time period in which producers may bring forth the supply of fish or other aquacultural products, supply is more elastic than in the short-run. This is a clear illustration of the condition shown in Figure 3-12.

Long-run supply is the quantity of product on the market as determined by the continuum of equilibria between changes in marginal revenue (MR) and LRMC. In a graphic sense, long-run supply is determined by the intersection of MR and the LRMC above its intersection with LRAC.

Economies of Scale

As the size of plant and the scale (intensity) of the operation increase, considering expansion from the smallest possible plant, certain economies of size or scale are usually realized. Over the range of output produced by Plants 1-3 in Figure 5-8, average costs are declining; these declining costs mean that costs are increasing less than proportionately with output. Since plant equilibria mini-

FIGURE 5-8. Long- and Short-Run Average Cost Curves, and Long- and Short-Run Marginal Cost Curves for a Firm in a Long-Run Equilibrium

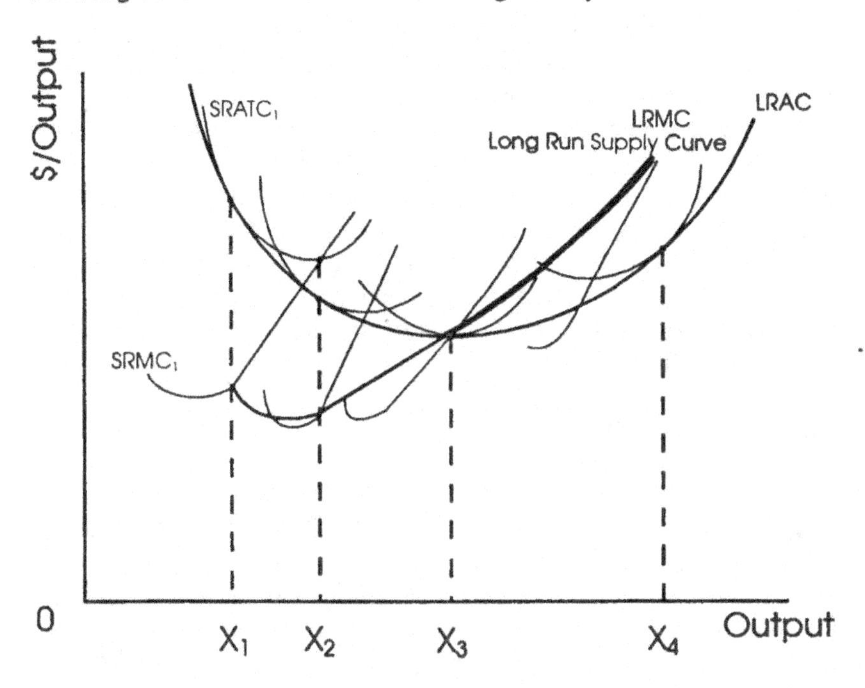

mum costs for 1 and 2 are greater than that for 4, the system exhibits decreasing returns to scale at this higher output. Figure 5-8 may also represent one firm with varying scales of operation. As production is intensified, the same results occur as with multiple numbers of plants.

Many factors combine to produce the pattern of increasing returns to scale. Economies, which cause long-run costs to decline, result from both production and market relationships. For example, labor productivity is usually greater when individuals are allowed to specialize. Specialization in the use of labor results in economies of scale. Similarly, there are market advantages to be gained when greater quantities are available from a single firm or geographical production area. Better prices may be realized through market concentration since both processing and distribution costs may be lowered.

Diseconomies of Scale

The rising portion of the long-run average cost curve is usually attributed to "diseconomies of scale," which means limitation to efficient management. Management of a fish processing plant entails managing a large number of workers. It entails careful supervision to attain a high quality product. As firm size increases and production requires added supervision, top management must delegate responsibility and authority to other employees. Contact with the daily routine of operation tends to be lost and efficiency of operation tends to decline. Just when diseconomies of scale begin to set in is hard to determine, but it does occur and its existence is rather significant in certain industries.

Firm Size and Plant Size

The long-run average cost curve could take three different shapes (Figure 5-9). If there are no economies or diseconomies of combining plants or intensifying daily operations, costs are said to be constant. Costs for scale or plant size might decline throughout the entire range of output. Such is the case of natural monopolies. Typically, however, costs first decline and then rise in the more familiar U-shaped fashion.

FIGURE 5-9. Economies of Scale or Size Under Alternative Conditions of Catfish Production

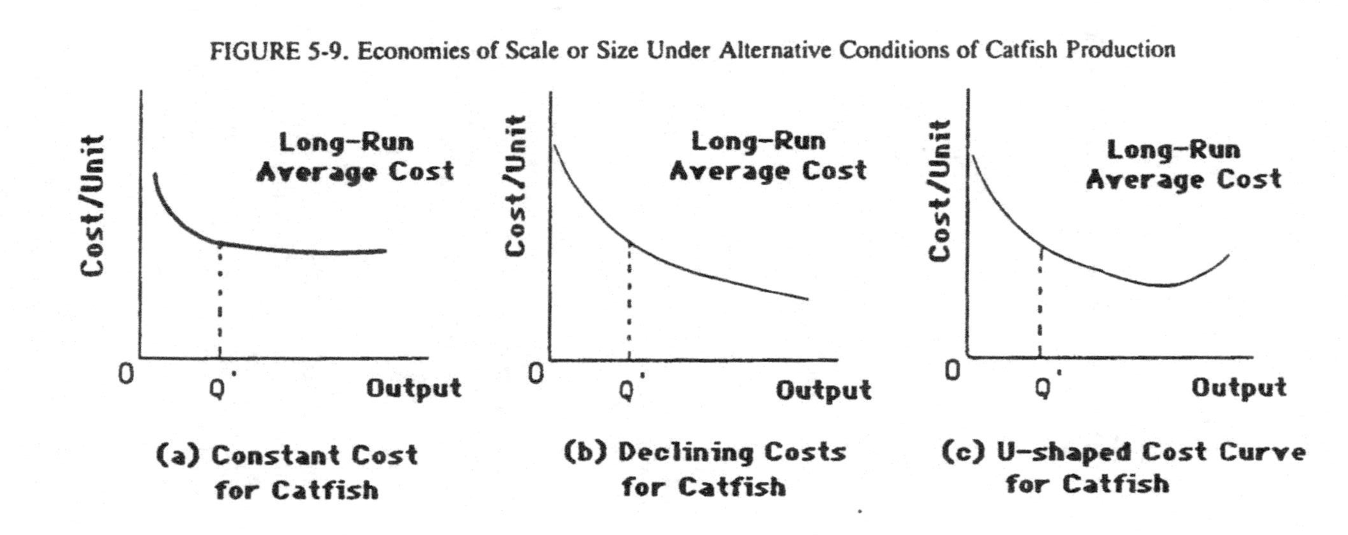

REFERENCES AND RECOMMENDED READINGS

Adams, C.M., W.L. Griffin, J.P. Nichols, and R.E. Brick. 1980. "Application of a Bio-Economic Engineering Model for Shrimp Mariculture Systems." *Southern Journal of Agricultural Economics*. July 135-141.

Allen, G.P., L.W. Botsford, A.M. Schuur, and W.E. Johnston. 1984. *Bioeconomics of Aquaculture*, New York: Elsevier. 351 pages.

Blommestein, E., H. Deese, and J.P. McVey. 1977. "Socio-Economic Feasibility of *Macrobrachium rosenbergii*, Farming in Palario." *Journal of World Maricultural Society*. pp. 747-63.

Browning, E.E. and J.M. Browning. 1976. *Micro-economic Theory and Applications*, Second Ed. Englewood Cliffs, NJ: Prentice-Hall.

Dellenbarger, L.E., L.R. Vandeveer, and T.M. Clarke. 1987. *Estimated Investment Requirements Production Costs and Breakeven Prices for Crawfish in Louisiana*. D.A.E. Research Report 670, Louisiana Agricultural Experiment Station, Baton Rouge, Louisiana. 37 pages.

Engle, C. R. 1988. "An Economic Comparison of Aeration Devices for Aquaculture Ponds." *Aquacultural Engineering*. (8)193-207.

Ferguson, C.E. and S.C. Maurice. 1974. *Economic Analysis*, Illinois: Richard D. Irvin, Inc.

Foster, T. H. and J. E. Waldrop. 1972. *Cost-Size Relationships in the Production of Pond-Raised Catfish for Food*, Bulletin 792, Mississippi State University, Agricultural and Forestry Experiment Station, 69 pages.

Gates, J.M., C.R. MacDonald, and B.J. Pollard. 1980. *Salmon Culture in Water Reuse Systems: An Economic Analysis*, Technical Report 78; University of Rhode Island. 52 pages.

Griffin, W., A. Lawrence, and M. John. 1984. "Economics of Penaeid Culture in the Americas." Proceedings of the First International Conference on the Culture of Penaeid Prawns/Shrimps, Iloilo City, Philippines, 151-60.

Panayotou, T., S. Wattanutchariya, S. Isvilanonda and R. Tokrisna. 1982. *The Economics of Catfish Farming in Central Thailand, 1982*. ICLARM Technical Reports, 4, Manila, Philippines. p. 60.

Pardy, R.C., W.L. Griffin, M.A. Johns, and A.L. Lawrence. 1983. "A Preliminary Economic Analysis of Stocking Strategies for Penaeid Shrimp Culture." *Journal of World Maricultural Society*. 14:49-63.

Shigekawa, K.J. and S.H. Logan. 1986. "Economic Analysis of Commercial Hatchery, Production of Sturgeon." *Aquaculture*. 51:299-312.

Wattanutchariya, S. and T. Panayotou. "The Economics of Aquaculture: The Case of Catfish in Thailand." Proceedings of a Workshop held in Singapore International Development Center, Ottawa, Ontario, Canada, pp. 26-34.

Chapter 6

Factor-Factor and Product-Product Relationships

HOW TO COMBINE INPUTS AND OUTPUTS

The use of one variable input (factor) in the production of a given level of fish while other inputs (factors) were held constant has been examined. Now, a study of the use of two variable inputs in the production of a given level of fish must be considered. There are a number of variable input combinations a farmer may use in the production of a given quantity of fish. The fish farmer tries to combine inputs in such a way as to maintain a constant level of production. The farmer is trying to find the correct combination of inputs that will generate the greatest amount of profit. Therefore, the farmer minimizes costs by using the resources that are less expensive and usually most abundant to substitute for more expensive resources. In general terms, this is called factor-factor substitution. In economic literature one finds the factor-factor relationship also referred to as the least cost principle. The resources used in the production of a product may be divided into two categories. One category includes those resources required to produce the product, but which remain constant and are independent of the level of use of other resources. A second category includes regional inputs which vary with output, but may be substitutable within some range with another input. Both types may vary with output; the first in direct proportion, the latter in variable proportion. For example, it has been shown that bone meal may be substituted for feather meal and coconut oil meal may be substituted for soybean oil meal to a certain level in the production of feed concentrate. Also, Fowler

(1980) found that cottonseed meal was an efficient replacement for fish meal in diets fed to chinook salmon.

The rising cost of protein feed and inorganic fertilizer, as well as the general concern for energy conservation, have stimulated increased interest in the utilization of animal manures as substitutes in aquaculture. Animal manures may replace inorganic fertilizer and, to a limited extent, commercial feeds depending on the aquacultural system.

Species such as tilapias have the ability to feed on plankton, but also feed on bottom materials (detritus). They also may be produced using artificial feed. Therefore, farm managers have produced tilapia by combining organic fertilizer and artificial feed. In areas such as the United States where artificial feed is relatively abundant, farmers rely heavily on feed. In developing countries where artificial feed is expensive, farmers increase the use of manure and compost in tilapia production.

Modern fish diets, both dry and moist, rely heavily on fish meal to supply the principal portion of their dietary protein (Fowler and Banks, 1970). In recent years, fish meal has become increasingly difficult to obtain. In 1972-1973, the Peruvian anchovy (*Engravlis ringens*) fishery crisis forced a search for other sources of protein for fish diets. Previously this species had supplied more than 80 percent of world production of saleable fish meal (Anon., 1973).

In experiments carried out at the Fish Farming Experiment Station in Stuttgart, Arkansas, the growth of channel catfish reared on two diets was evaluated. One diet contained fish meal and the other a commercially available fish meal substitute to determine if the substitute would be a suitable alternative to fish meal. It was found that the cheaper fish substitute replaced the fish meal. The diets contained 10 percent fish meal or 10 percent of the substitute.

Cacho (1988) stated that the choice of a feeding strategy in catfish production represented two decision variables of different natures, but closely related, namely diet quality and diet quantity. He further stated that there was some degree of substitutability between diet quality and feeding rates in their effect on growth rates.

The factor-factor model answers the question of how much of two inputs, X_1 and X_2, should be combined to produce a given level of output so that the objective of cost minimization (maximum economic efficiency) might be realized. Every input combination will have a unique technical efficiency (rate of gain and costs). A manager, or fish farmer, must evaluate the many possibilities to find the one giving the greatest profit. Figure 6-1 provides an illustration of two variable inputs which are used in the production of different quantities of fish. Each level, denoted by curves AA, BB, and CC, represents a particular level of total production which is possible by varying the quantities of each input. Different input combinations may produce the same output because they are substitutable. The level of output AA, BB, etc., is called an isoquant, meaning equal quantity at each input combination.

The *isoquant curve* illustrates the possible combination of inputs that will yield a given amount of output. Figure 6-2 shows the level of two inputs, X_1 and X_2, which may be used to produce 1,500 kilograms (3300 pounds) of catfish. See segment CC in Figure 6-1. The isoquant has a negative slope, which reflects a diminishing marginal rate of technical substitution (MRTS).

The *marginal rate of technical substitution* (MRTS) refers to the

FIGURE 6-1. Output as Function of Two Variable Input

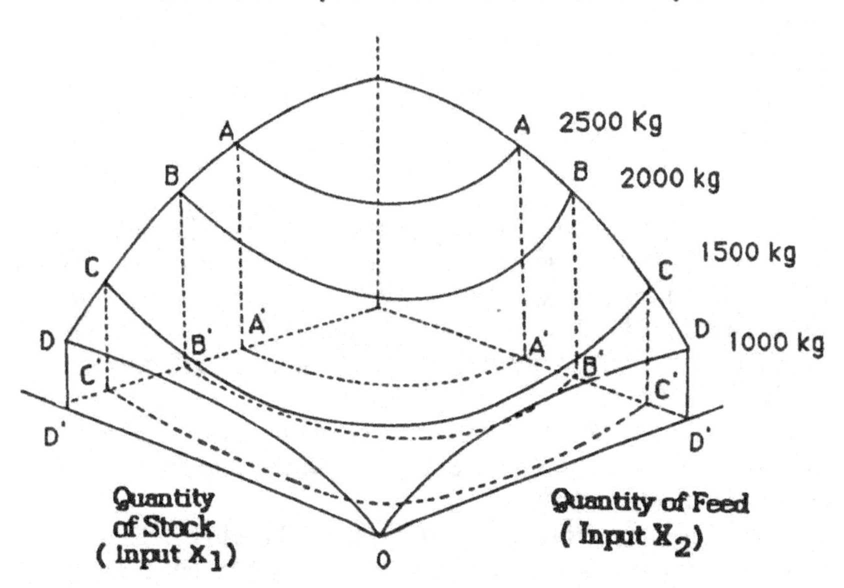

FIGURE 6-2. The Hypothetical Isoquant for 1500 Kilograms (3300 Pounds) of Catfish

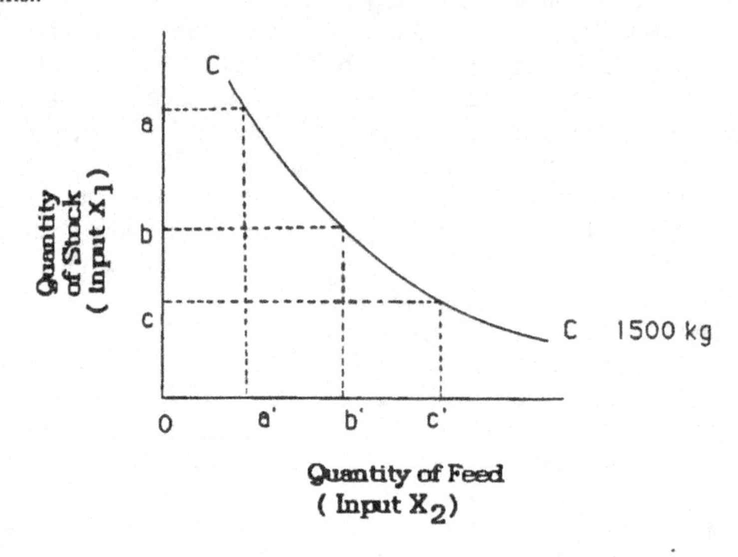

amount of a resource that may be decreased as use of another resource is increased by one unit without affecting output. Algebraically stated:

$$\text{MRTS } X_2 \text{ for } X_1 = \frac{\Delta X_1}{\Delta X_2}$$

In Figure 6-2, if one wants to produce 1,500 kilograms (3300 pounds) of fish there is a choice of several combinations of inputs X_1 and X_2. If the level of input X_1 is increased, the level of input X_2 must be reduced. The replacement of input X_1 by X_2 without changing the level of output gives some indiction of how effective X_1 can replace X_2.

The marginal rate of substitution is almost always stated as a negative; the iso-product curve slopes downward and to the right in a two-dimensional drawing.

A diminishing MRTS is apparent when successive equal units of a variable input (X_2) are substituted for another variable input (X_1)

and the successive equal units of the substitute (X_2) gradually replace less and less of the original value of input (X_1).

In Table 6-1, the first substitution of X_2 for X_1, moving from 4 to 5 units of X_2, 1 unit of X_2 replaced 4.0 units of X_1; changing from 5 to 6 units of X_2 replaced 2.0 units of X_1. With the third substitution, 6 to 7 units of X_2, 1 unit of X_2 replaced 1.5 units of X_1, and with the last substitution, 7 to 8 units of X_2, 1 unit of X_2 replaced 0.5 units of X_1. As more X_2 was used in producing 3,000 units of Y, X_2 gradually became more productive relative to X_1. This process could be reversed by substituting one unit increment of X_1 for X_2 and observing the same characteristics.

Returning to feed quality and feeding rate, Cacho (1988) stated that increasing feeding rate of a given diet or increasing dietary protein content for a given feeding rate can increase growth rate, and both strategies will decrease water quality which in turn affects fish appetite, feed efficiency, and growth rate. Cole and Boyd (1986) reported that water quality deteriorated proportionally as feeding rate was increased in catfish ponds, while generation requirements (to keep the fish alive) increased at an increasing rate.

Cacho et al. (1989) showed different combinations of protein and feed that produced a given weight of fish in a fixed time. Figure 6-3 shows that in the production of a given weight fish, as dietary protein is decreased, the quantity of feed must be increased to produce the same level of output. One would expect the isoquant to shift to the right as the desired fish weight is increased. The shape of the isoquant indicates the degree of substitutability between dietary protein and quantity of feed consumed.

TABLE 6-1. The Marginal Rate of Substitution of Inputs X_1 and X_2 in the Production of 1500 Kilograms (3300 Pounds) of Fish

Units of X_2	Units of X_1	Change in inputs ΔX_2	$\Delta X1$	MRTS: X_2 for X_1 $\Delta X_1/\Delta X_2$	MRTS: X_1 for X_2 $\Delta X_2/\Delta X_1$
4	10				
5	6	1	-4	-4/1 = -4.0	-1/4 = -0.25
6	4	1	-2	-2/1 = -2.0	-1/2 = -0.50
7	2.5	1	-1.5	-1.5/1 = -1.5	-1/1.5 = -0.67
8	2.0	1	-0.5	-0.5/1 = -0.5	-1/5 = -2.00

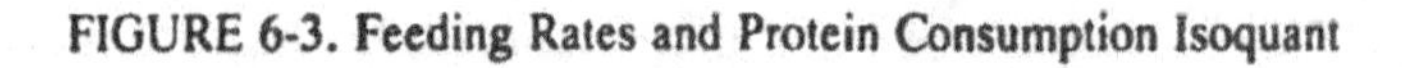

FIGURE 6-3. Feeding Rates and Protein Consumption Isoquant

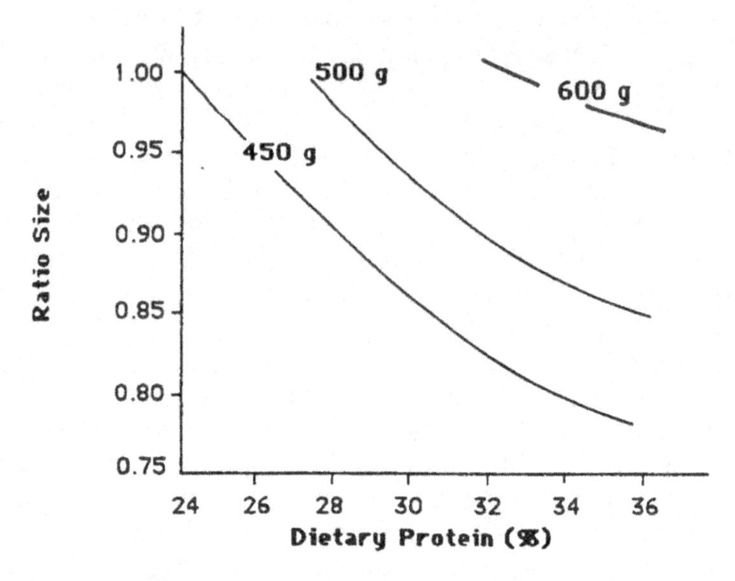

Source: Cacho, O. J., U. Hatch and H. Kinnucan (1990), Bioeconomics Analysis of Fish Growth: Effects of Dietary Protein and Ration Size. *Aquaculture*, 233-238.

Relation of MRTS to Marginal Products

The MRTS is related to the marginal products of the two inputs. If two inputs, X_1 representing units of feed quantity (feeding rate), and X_2 representing protein quality, are used in the production of a given quantity of tilapia, the marginal productivity (MP) of feed quantity will be higher as more protein quality is used in production. This will be so up to a point where the MP for feed will decline. The same is true for feeding rate.

The movement along the isoquant is represented by ΔQ which is equal to 0. This is so since output does not increase when moving along the isoquant. The total change may therefore be represented by the equation:

$$\frac{MP_{x_1}}{MP_{x_2}} = -\frac{\Delta X_2}{\Delta X_1}$$

As noted earlier, MRTS deals with two different inputs while holding output and other inputs constant. Taking the formula for marginal product fo each of the inputs:

$$MP_{x1} = \frac{\Delta Q_1}{\Delta X_1} \text{ and } MP_{x2} = \frac{\Delta Q_2}{\Delta X_2}$$

and solving each equation for ΔQ gives:

$$\Delta Q = \Delta X_1 MP_{x_1} \text{ and } \Delta Q = \Delta X_2 MP_{x2}$$

since:

$$\Delta Q = 0$$

Then: $\Delta X_1 (MP_{x_1} = \Delta X_2 (MP_{x_2})$.

If each side is divided by:

$$\Delta X_2 (MP_{x_2})$$

the following equation is derived:

$$\frac{MP_{x_1}}{MP_{x_2}} = -\frac{\Delta X_2}{\Delta X_1} = MRTS_{x_2} \text{ for } X_1$$

The MRTS at a point on an isoquant is, therefore, equal to the ratio of the MPs of the two inputs. In this case, the MRTS of feeding rate for protein quality in the production of a given quantity of fish is the marginal product of protein quality divided by the marginal product of feed (feeding rate).

The Isocost Curve

The fish farmer normally has a fixed budget. If so much feed is purchased, only limited purchases of other inputs are possible. For example, if the operating budget for producing an acre of catfish is \$200, and \$120 worth of feed are bought, only \$80 of the other required inputs for the production of an acre of catfish may be purchased. If the farmer buys \$100 worth of feed, \$100 of other inputs may be bought.

The *isocost curve* represents the combination of two inputs, X_1 and X_2, that may be purchased with a given budget. As stated, if the farmer desires to purchase more feed without spending more money, other input purchases must be reduced. The amount of each that my be bought depends upon the price of each and the money available.

Assume \$8 is available to purchase X_1 and X_2. If $P_{x1} = \$1$, and $P_{x2} = \$2$, then 4 units of X_2 and 8 of X_1 could be bought. If the \$8 were used to buy both X_1 and X_2, then any amount of each whose combined cost was \$8 could be bought. With such information, one may draw the isocost line by connecting the two points, $X_1 = 8$ or $X_2 = 4$. If prices doubled, 4 units of X_1 and 2 units of X_2 could be purchased. Any point on the \$8 line in Figure 6-4 represents the amount that my be bought for \$8. The slope of a line is defined as $\Delta Y_1/\Delta X_1$, or simply stated in less technical terms, rise over run. In

FIGURE 6-4. Isocost Line Illustrated

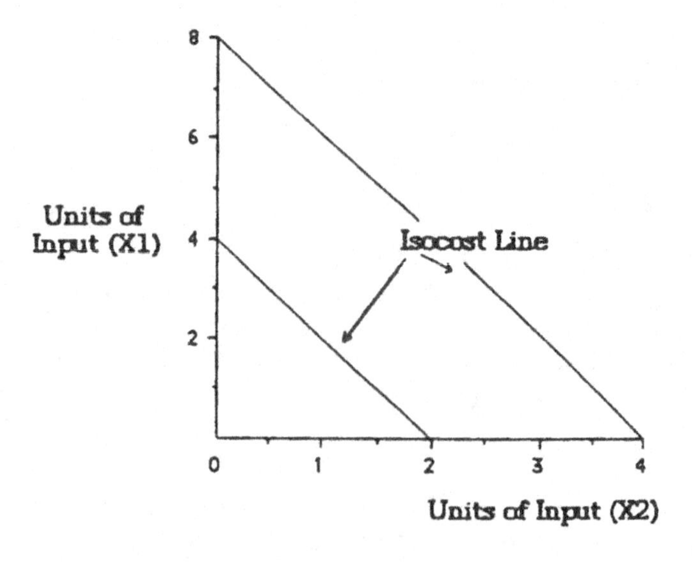

this case, $\Delta X_1/\Delta X_2$. Since the isocost curve is a straight line, the slope is constant and a large change may be viewed. Actually, it is easier to consider an all-or-nothing situation. If all of the budget is used on X_1, then B/P_{x1} can be purchased. Conversely, if all of the budget is spent on X_2, then B/P_{x2} of X_2 may be purchased. Thus, two points on the isocost curve have been defined $(B/P_{x1},0)$ and $(0,B/P_{x2})$. Thus the slope is:

$$\frac{B/P_{x1} - 0}{0 - B/P_{x2}} \text{ or } - \frac{B/P_x}{B/P_x} \text{ which is } - \frac{P_{x1}}{P_{x2}}$$

In other words, the slope of the isocost line is determined by the price ratios. Any changes in the price results in a different slope.

PROFIT MAXIMIZATION

Maximizing profit requires that the product be produced at minimum cost. To determine the least cost input combination, the price of each input and the MRTS between inputs must be known. When they are known, the profit maximizing combination may be determined by finding the point where the MRTS is equal to the inverse of price ratios. This point is synonymous with the point where the isoquant is tangent to the price ratio (see Figure 6-5).

The MP of each unit of input should be related to its price. A fish farmer who uses any two inputs, say labor and capital, in the production of tilapia should use each of the inputs so that the MP of one input divided by its price is equal to the MP of the other input divided by its price. Assuming X_1 represents labor and X_2 represents capital, and $MP_{x1} = 20$ and $MP_{x2} = 40$, the price of $X_1 = \$2.00$ and the price of $X_2 = \$4.00$, it would seem logical that the fish farmer would use quantities of each input in the production of a given quantity of fish to the point where their MPs per dollar are equal. In this case:

$$\frac{MPx_1}{Px_1} = \frac{20}{2} = 10 \text{ and } \frac{MPx_2}{Px_2} = \frac{40}{4} = 10$$

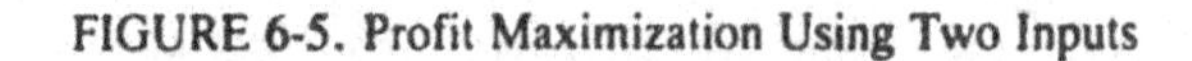

FIGURE 6-5. Profit Maximization Using Two Inputs

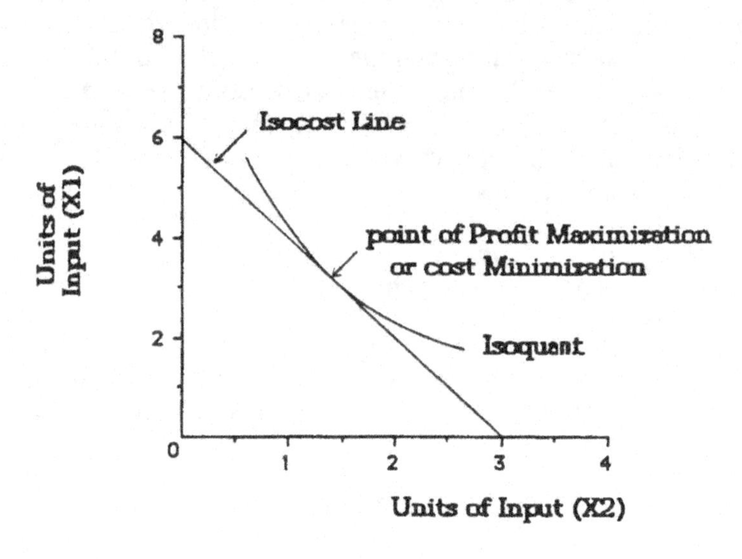

At this point, the fish farmer has no incentive to replace one input with the other since their marginal products in relation to their prices are equal. This relationship would hold for any number of inputs.

$$\frac{MP_{X_1}}{P_{X_1}} = \frac{MP_{X_2}}{P_{X_2}} = \cdots \frac{MP_{X_n}}{P_{X_n}}$$

In Figure 6-6, the tangency of the isocost line to the isoquant at point B indicates that the input price ratio P_{X1}/P_{X2} is equal to the $MRTS_{X2}$ for X_1.

It has been shown that:

$$MRTS_{X_1} \text{ for } X_2 = \frac{MP_{X_1}}{MP_{X_2}}$$

and that a farmer will choose a combination of inputs in the production of a given quantity of output where:

$$\frac{MP_{x_1}}{P_{x_1}} = \frac{MP_{x_2}}{P_{x_2}}$$

Therefore, it may be seen from the above equation that:

$$\frac{MP_{x_1}}{MP_{x_2}} = \frac{P_{x_1}}{P_{x_2}}$$

If a combination of X_1 and X_2 units of inputs were chosen at point A, the same quantity of output would be produced as at point B, but at a higher cost. Similarly, a choice of input combination at point C would produce the same level of output as at point B, but at a higher cost. Point B shows the point of least cost for the production of the given level of output.

Using Figure 6-3 and adding cost curves as illustrated in Figure 6-7, it is shown that the cost of producing a market-size fish may be minimized by finding the tangency point between the isoquant and the isocost curves. The optimum point is found at U (Figure 6-7), where the isocost touches the isoquant.

FIGURE 6-6. Point of Cost Minimization or Profit Maximization

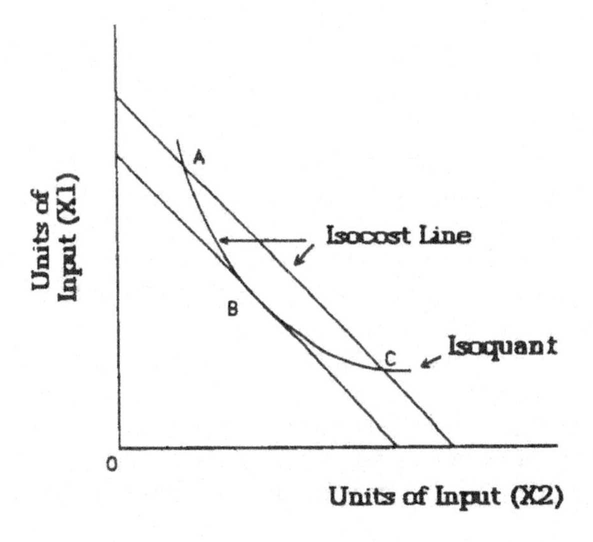

FIGURE 6-7. Cost Minimization for Feeding Rate and Protein Quality

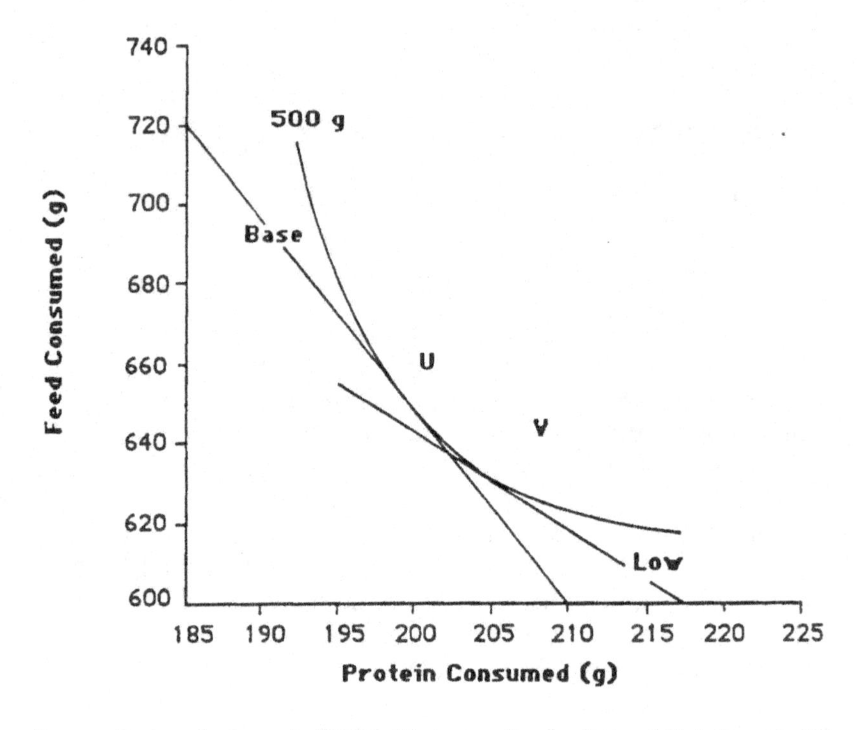

Source: Cacho, O. J. et al. (1990), Bioeconomics Analysis of Fish Growth: Effects of Dietary Protein and Ration Size. *Aquaculture*, 88, 234.

IMPERFECT SUBSTITUTION

Because of diminishing marginal productivity of resources, the usual resource combination situation is one in which MRTS is diminishing (Figure 6-8).

The shapes of the isoquants reveal a great deal about the substitutability of the input factors: that is, the ability to substitute one input for another in the production process. The slope of the isoquant provides the key to the substitutability of input factors.

With imperfect substitutes, careful consideration must be given to both the price relationship and the marginal rate of substitution. Profits may be increased whenever the substitute input's price relative to the other input price is less than the physical rate at which the

FIGURE 6-8. Imperfect Substitution

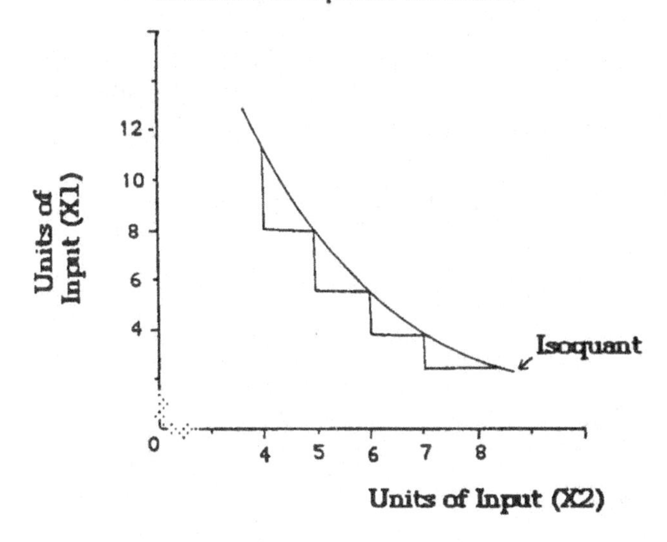

farmer substitutes for the latter. If X_2, the substitute, is twice as costly as X_1, but it is three times as productive, cost will be reduced by substituting X_2 for X_1.

Constant Substitution

There are two cases of constant substitution between two variable resources. When one unit of a resource may be substituted for one unit of another resource without changing output, the two are *perfect substitutes*. For example, two fertilizers of different brands but with identical contents are fully substitutable. Figure 6-9 illustrates perfect substitutes. The other case of constant substitutability is the situation when two inputs substitute at a constant rate 1:2 or 1:3 or 1:4, shown in Figure 6-10.

Profit Maximization for Constant Substitutes

Determining the least cost combination is easier for constant substitutes than for imperfect substitutes. From a cost standpoint, fish farmers will use one input, the least expensive one. They will maximize profits and there is no incentive to combine the two inputs. If the prices are the same, farmers will be indifferent. If the prices are different, farmers will choose the least expensive item.

Perfect Complementarity

At times the aquaculturist faces a situation of no substitutability. When using a tractor, a driver is needed. Such a situation is called perfect complementarity and the MRTS = 0 (Figure 6-11). Cacho

FIGURE 6-9. Perfect Substitution

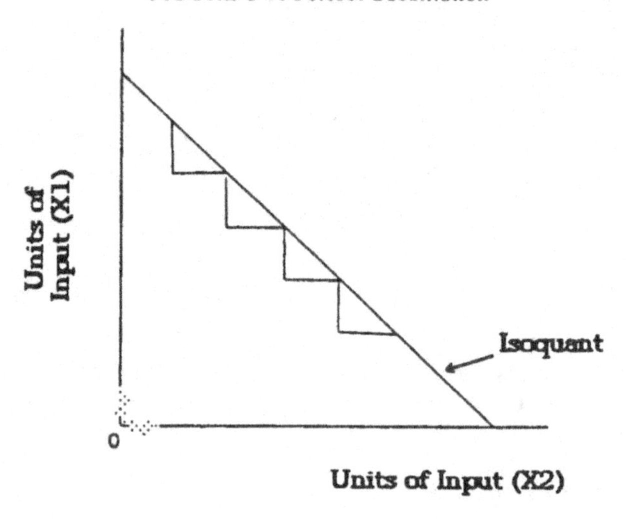

FIGURE 6-10. Constant Substitution

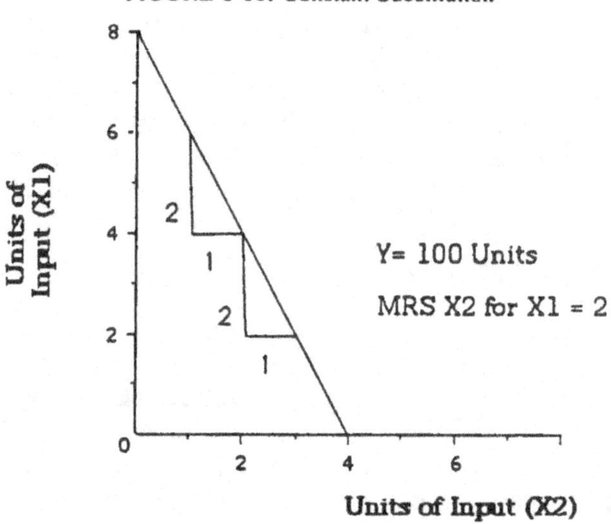

FIGURE 6-11. Perfect Complements

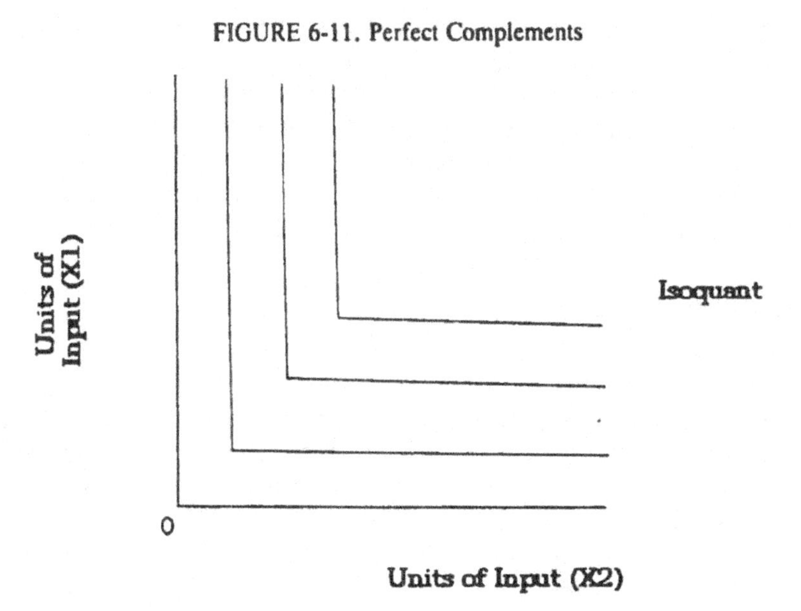

indicated that the feed and energy (in the form of aeration) are complementary inputs, and the relationship is stronger as feeding rates increase. Perfect complements may be represented in isoquant form as two straight lines parallel to the axes and joined at right angles, implying that increases in one factor will not increase output unless the other (complementary) factor is also increased. This is particularly so for feed and aeration at high feeding rates or high stocking densities.

A special case arises when inputs must be used in *fixed* proportions to produce the product. If one input is purchased, the other must be purchased in exact proportions irrespective of the price.

Expansion Path

To this point a single optimal combination of inputs to produce a given level of output has been discussed. The expansion path extends this concept to show the optimal combination of inputs over a range of output levels. The optimal level of production, or the least cost combination of resources to attain a given level of production, will change with movements to higher or lower isoquants (levels of output). In Figure 6-12 DD>CC>BB>AA. The points of tangency, as output is varied, describe a line called the expansion path. This line represents the optimal combination of the two resources to achieve successively higher levels of output. In the long-run, the farmer who intends to expand operations will follow this path if intending to minimize cost, i.e., to achieve maximum economic production efficiency.

Economic Region of Production

Over a range, the isoquants are negatively sloped. The parallel lines in Figure 6-13 indicate the points at which the isoquants bend back upon themselves. The lines 0K and 0L join these points and form the boundaries fo the rational regions of production. The boundary lines indicate the points where the $MRTS_{x1}$ for X_2 (or X_2 for X_1) is 0. These are called ridge lines since they connect the points where the isoquants "bend back" or reverse their slope reduction.

FIGURE 6-12. Illustration of the Expansion Path Following Least Cost Combination of Two Inputs

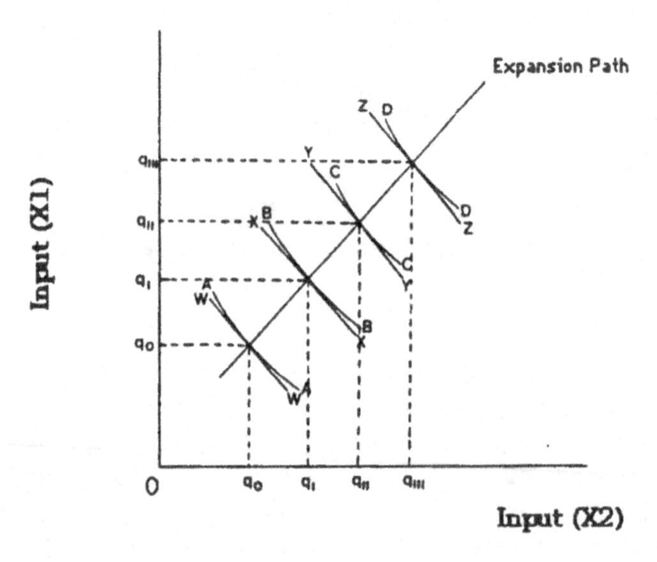

FIGURE 6-13. The Rational Boundary of Production and Ridge Lines

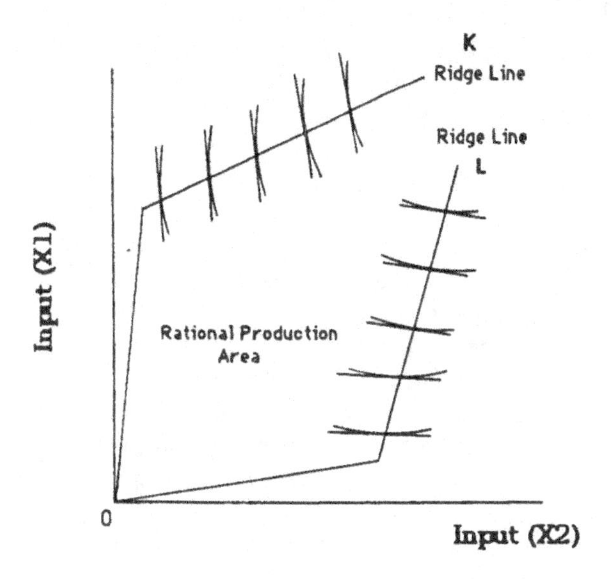

THE PRODUCTION OF TWO OUTPUTS

Fish producers have learned to maximize the level of output by producing species which complement each other. If the species are chosen carefully and in the right proportion, it has been observed that total production of the two species is increased. However, if species are not carefully chosen, the total output may be lessened. For example, consider a predator/prey species mix. Too few prey will depress growth of the predator from insufficient food as well as from lowered success in finding prey. Too many prey will prevent the predator from controlling reproduction and prey population growth will be excessive.

In polyculture, the rule is to mix species that complement rather than compete with each other. Hence, carp species that consume different types of food are chosen in a polyculture system. Thus, in the process of optimizing total pond productivity, the farmer should consider different ecological niches at different strata of water column by multi-species fish (Alsagoff et al., 1990). A classic example in theory is the Chinese carp polyculture system which combines upper depth inhabiting macrophytic herbivores with mid-depth planktonic herbivores as well as bottom omnivores. In fact, ecological complementarity is further maximized by selecting such species (e.g., the common carp consuming fecal wastes of the grass carp [Bardach et al., 1972]).

Product-Product Substitution

The product-product model attempts to answer the question of what products the firm should produce so as to satisfy the objective of maximizing returns for a given level of cost. Two microeconomic tools are necessary for solving the problem posed in this model. The first is the product transformation curve, also known as the production possibility curve. The curve, illustrated in Figure 6-14, shows the possible combinations of products Y_1 and Y_2 that can be produced with a given stock of input and technology. The second is the isorevenue curve.

We may produce either all of Y_1, all of Y_2, or a combination of

FIGURE 6-14. Return Maximizing Combination of Products Y_1 and Y_2

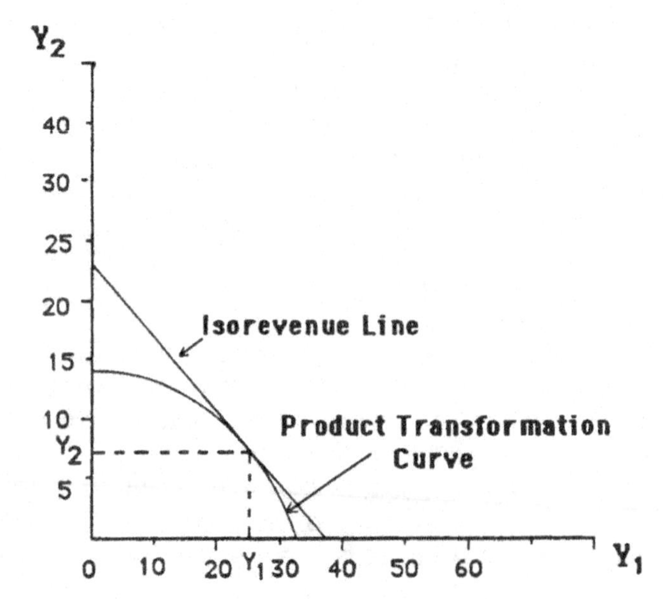

both. If we increase the production of Y_2, then more and more of Y_1 must be sacrificed and vice versa. The rate at which one output is sacrificed for the other is called marginal rate of product substitution (MRPS).

The Marginal Rate of Product Substitution

The marginal rate of product substitution (MRPS) is the amount by which an output changes in quantity when the other output is increased by one unit, given that the amount of input and technology used remain constant. The MRPS of Y_1 for Y_2 is:

$$MRPS = \frac{\Delta Y_2}{\Delta Y_1}$$

The MRPS represents the slope of the production possibility curve.

The Isorevenue Line

This function shows the combination of the two products, Y_1 and Y_2, that will generate a given level of revenue. The slope of the isorevenue line, also shown in Figure 6-14, is constant throughout and is dictated by the prices of the products just as the isocost line was dictated by factor prices. The slope of the isorevenue line is:

$$-\frac{\Delta PY_1}{\Delta PY_2}$$

Maximum Return Combination

The production possibility curve represents all possible combinations of two products that could be produced using a given amount of variable input. Only one combination of output will be produced in practice. Since the goal is to move the isorevenue line as far from the origin as possible, the return maximizing combination of Y_1 and Y_2 would occur where the two curves are tangent. At this point, the slope of the product transformation curve is equal to the slope fo the isorevenue line. Figure 6-14 shows the point of maximum combination of output which is where:

$$\frac{\Delta Y_2}{\Delta Y_1} = -\frac{P_{y_1}}{P_{y_2}}$$

$$\frac{MPP\ Y_2}{MPP\ Y_1} = \frac{P_{y_1}}{P_{y_2}} \text{ or } MPP\ Y_2 * P_{y_2} = MPP\ Y_1 * P_{y_1}$$

Competitive Products

Products are termed competitive when an increase in one product may come about only by reducing the output of other products. Outputs are competitive because they require the same input. This situation is illustrated by Figure 6-15(A). The productions of two species which have the same feeding habits is likely to cause competition for feed at a given point in the production process. However, Van Dam (1990) found that at certain stocking densities, even rice and fish competed to cause reduced yield.

Complementary Products

Two products are complementary if an increase in one product causes an increase in the second product when the total amount of inputs used on the two are held constant, illustrated by Figure 6-15(B). The production of rice and fish may be considered complementary to a certain extent. Yet, as seen above, they may be competitive. The same is true for many product combinations. There may be a restricted quantity range where they either complement or compete with one another.

Supplementary Products

Two products are supplementary if the amount of one may be increased without increasing or decreasing the amount of the other, illustrated by Figure 6-15(C). The production of two species in a pond with different feed requirements may be supplementary at a given point in the production process. An example is the production of carnivore and herbivore species.

FIGURE 6-15. Production Possibility Curves Showing Possible Relationships Among Enterprises

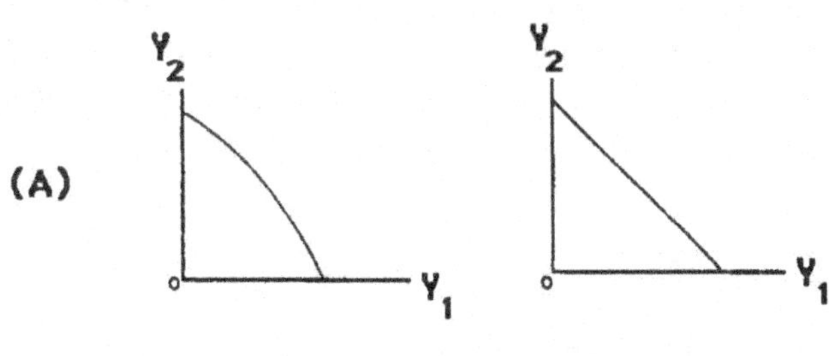

B. COMPLEMENTARY

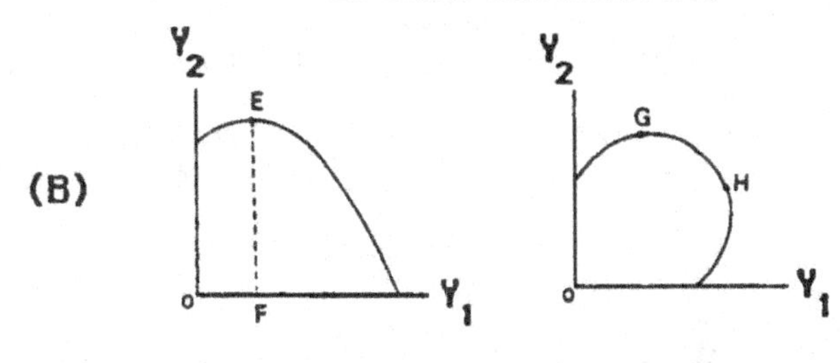

C. SUPPLEMENTARY

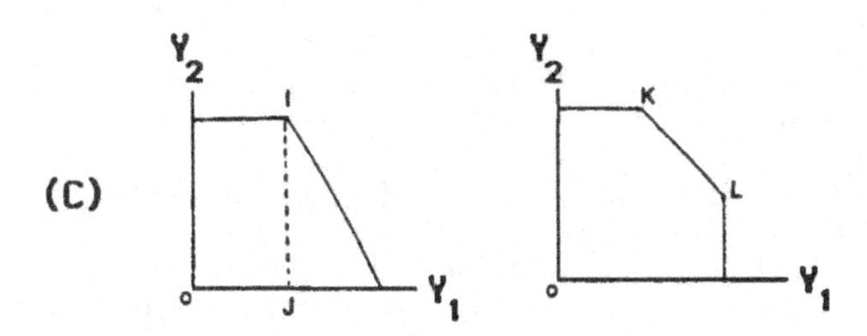

D. JOINT

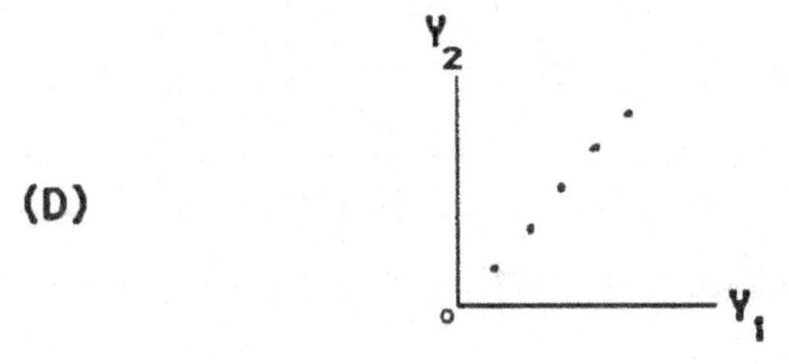

Joint Products

Products which result from the same production process are termed joint products. As a rule, the two are combined and production of one without the other is impossible. An illustration is shown in Figure 6-15(D). The production of mother-of-pearl and pearls may be considered as joint products.

REFERENCES AND RECOMMENDED READINGS

Alsagoff, S.A.K., H.A. Clonts, and C.M. Jolly. 1990. "An Integrated Poultry, Multi-Species Aquaculture for Malaysian Rice Farmers: A Mixed Integer Programming Approach." *Agricultural Systems.* 32:207-31.

Anonymous. 1973. "What's Happening to Fish Meal?" *American Fish/U.S. Trout News*, 18(6):6-9.

Banks, J.L. and E.G. Fowler. 1976. "Animal and Vegetable Substitutes for Fish Meal in the Abernathy Diet, 1973." *The Progressive Fish Culturist.* 38(3):123-6.

Bardach, J.E., J.H. Ryther, and W. McLarney. 1972. *Aquaculture, the Farming and Husbandry of Freshwater and Marine Organisms*, NY: John Wiley and Sons, Inc. 868 pages.

Bishop, C. E. and W. D. Toussaint. 1958. *Agricultural Economic Analysis*, New York: John Wiley and Sons, Inc., 258 pages.

Cacho, O., U. Hatch and H. Kinnucan. 1990. "Bioeconomic Analysis of Fish Growth: Effects of Dietary Protein and Ration Size." 233-238.

Cacho, J.O. 1988. *A 1988 Bioeconomic Model for Catfish Production with Emphasis on Nutritional Aspects*. Ph.D dissertation, Auburn University, Auburn, Alabama.

Cacho, O., U. Hatch, and H. Kinnucan. 1989. *Bioeconomic Optimization of Catfish Feeding*, Working paper 89-3, Department of Agriculture Economics and Rural Sociology, Auburn University, Auburn, Alabama.

Chien, Y. and J.W. Avault, Jr. 1980. "Production of Crayfish in Rice Fields." *The Progressive Fish-Culturist*. 42(2):67-71.

Clonts, H.A., C.M. Jolly, and S.A. Alsagoff. 1989. "An Ecological Foodniche Concept as a Proxy of Fish-pond Stocking Rates in a Mixed Integer Linear. Programming Model for Integrated Aquacultural Farming for Malaysia." *Journal of the World Aquaculture Society*. 20(4):268-76.

Cole, B. A. and C.E. Boyd. 1986. "Feeding Rate, Water Quality, and Channel Catfish Production in Ponds." *Progressive Fish Culture*. 48:25-29.

Fowler, L.G. 1980. "Substitution of Soybean and Cottonseed Products for Fish Meal in Diets Fed to Chinook and Coho Salmon." *The Progressive Fish Culturist*. 42(2):87-91.

Fowler, L.G. and J.L. Banks. 1970. "Test of Substitute Ingredients and Effects of Storage in The Abernathy Salmon Diet in Bureau of Sport Fisheries and Wildlife." (1968) Technical Paper. No. 47, p. 8.

Greenland, D.C., H.K. Dupree, and C.H. Long. 1984. "Growth Studies Conducted on Channel Catfish Fed Diets with Fish Meal or Fish Meal Substitute." *Feedstuffs*, p. 20.

Lizama, L.C., L.R. McDowell, and J.E. Marion. 1989. "Utilization of Aquatic Plants. *Elodea canadensis* and *Hydrilla verticillata* in Laying Hen Diets." *Nutrition Reports International*. 39(3):521-36.

Moshen, A.A. 1988. *Substituting Animal Protein Sources Into Corn-Soybean Meal*, an unpublished M.S. Thesis, Auburn University, Auburn, Alabama, 30 pg.

Pullin, R.S. and A.H. Shehadeh. 1980. "Integrated Agriculture-Aquaculture Farming Systems." Proceedings: ICLARM-SEARCA Conference on Integrated Agriculture-Aquaculture, Farming Systems, Manila, Philippines, 1979:195-208.

Rouse, D.B., G.O. El Naggar, and M.A. Mulla. 1987. "Effects of Stocking Size and Density of Tilapia on *Macrobrachium rosenbergii* in Polyculture." *Journal of the World Aquaculture Society*. 18(2):57-60.

Schroeder, G. 1986. "Integrated Fish Farming: An International Effort." *American Journal of Alternative Agriculture*, 1(3):127-30.

Smith, I.R. 1972. "Microeconomics of Existing Aquaculture Production Systems: Basic Concepts and Definitions." *Aquaculture Economics Research in Asia*. Proceedings of a Workshop in Singapore, June.

Van Dam, A.A. 1990. "Multiple Regression Analysis of Accumulated Data From Aquaculture Experiments: A Rice-Fish Culture Example." *Aquaculture and Fisheries Management*. 21:1-15.

Wilson, R.P. and W.E. Poe. 1985. "Effects of Feeding Soybean Meal with Varying Trypsin Inhibitor Activities on Growth of Fingerling Channel Catfish." *Aquaculture*. pp. (46)19-25.

Chapter 7

Farm Management

Rapid population growth and changes in consumer tastes have increased the demand for fish and moved the price upward. The change in demand has, in turn, forced nations to search for alternative sources of protein. A simultaneous increase in aquacultural production has occurred, but it is still insufficient to meet world demand. In the 6-year period between 1975 to 1981, fish production through aquaculture increased 43 percent, but in the following 6 years, production rose only about 20 percent. In 1986, FAO estimates showed current production of aquacultural fish species to be 10.21 million metric (10.4 million U.S.) tons.

The sudden increase in aquacultural production has been accompanied by an alarming number of aquacultural project failures. Consequently, more emphasis is now being placed on management of such enterprises with a goal of reversing the production-consumption squeeze worldwide.

Management of any farm involves both planning and operation. If planning is overlooked, the business will likely fail. Thus, the first farming decisions must be planning decisions, many of which are economic.

FARM PLANNING

Farm planning means assessing the implications of allocating resources in a particular way before making production decisions. It is an essential part of the rational decision process. A farm plan is an outline or scheme for the organization and utilization of the resources available on a given farm. Operating without such a plan means making decisions at random. All farm managers have a farm plan, but not all are written. On larger commercial farms, it is likely that formal procedures and detailed outlines are written. However, farmers who operate small farms all too often have plans in their heads. This is not good practice since the farm manager usually needs a plan with formal budgeting procedures to have closer control of the farm business, and for frequent economic problems such as persuading credit agents to provide loans.

Farm planning techniques are designed for use on individual farms, taking into account constraints of the resource endowment as well as any goals and objectives of the farm household. There are many techniques used in farm planning, but the most common are whole farm and partial budgeting and financial analysis.

Given a goal of profit maximization, the objective of farm planning and budgeting is to find the most profitable organization of the available resources and estimate the potential profit. There may be many alternative ways to combine the resources. Hence, the objective in planning is to systematically eliminate the less profitable enterprises or activities while working with the most profitable. The profit for the final plan is summarized in the whole farm budget, which may be compared with the profit from the current or alternative plans.

Whole farm planning and budgeting, like the other types of budgeting, is a technique for forward planning. The end product is a plan for the future. A whole farm plan and budget should be completed whether taking over an existing operation or starting a new farm business. A new plan may also be needed whenever a major change is considered.

DEVELOPING A FARM PLAN: ESSENTIAL STEPS

The Resource Inventory

The first step in developing the farm plan is taking an inventory of the available resources on the farm. Development of a *good* plan is directly dependent upon an accurate and complete resource inventory. These resources provide the means for production and profit, but they also place an upper boundary on potential production and profit.

Various resources to be considered in a plan are land, labor, capital, and management, the basic factors of production. The quantity and quality of land including water resources and atmospheric conditions have already been discussed. Labor implies both skilled and unskilled labor resources. Capital includes physical equipment, etc., and cash on hand or accessible through credit sources. Management refers to the skills of the operator to combine resources so as to maximize profit or achieve other goals such as family welfare, resource conservation, or personal satisfaction.

Site Selection

Site selection must be made early in the fish farm planning process, especially in aquacultural production. Aquaculture site selection hinges on several related factors, including topography, hydrology, and soil conditions.

Topography

The fish farming site should be relatively flat. A slope of 5 percent or less is desirable if diked ponds are used. Slightly steeper slopes are acceptable if watershed ponds are employed. However, areas subject to flooding are undesirable pond sites.

Hydrology

Water should be adequate in quantity and quality. Year-round water availability is important in pond management. Replacement of about 90 percent of the water lost from seepage and evapotranspiration has been shown to be economical (Clonts and Williams, 1983). Different organisms require different mediums for growth. For example, rearing jumbo tiger shrimp (*Penaeus monodon*) in waters with salinity of 30 ppt or over instead of a brackish condition of 10-25 ppt will not be conducive to the type of culture desired. The water quality required for each species produced must be seriously evaluated before engaging in aquaculture (Rabanal, 1987).

Soil Characteristics

Pond soils must have water retention capacity and high fertility. Fertility is important in encouraging development of a benthic community. A technical analysis for acidity and the presence of toxic materials in the soil is also essential.

Type of Pond

The type of pond selected is determined by biological, economic, financial, and sociological conditions. Miyamura and Katoh (1986) suggested that ponds be classified by aquacultural or water supply conditions. For example:

1. Aquacultural type – relative (intensive or extensive) labor/capital/management mix
2. Water supply and discharge method – natural or motor power
3. Level of investment – capital intensive or low input aquaculture.

The aquacultural type depends directly on the size of the pond, the availability of resources for construction and the social and national particularities of products to be consumed. Water supply and discharge is a function of topography, soils, and socio-economic infrastructure.

Land Use

The use of aquacultural pond lands prior to pond development should be considered also. Prior uses may leave residual materials either favorable or detrimental to aquatic species.

If the land is to be purchased, one must ascertain that aquacultural production will be the highest and best land use. If it is not, aquacultural activities will result in less than optimal gains (losses) to the operator. The owner of the aquacultural enterprise must be able to pay the land costs whether land is purchased or leased.[1]

Other factors which might affect land use include accessibility and distance from market. Is it physically possible for vehicles to go to and from the farm? How far is the farm from the market? Can fish be transported to and from the market without major investments? Are the roads accessible year round? Is the pond close to an urban area? There are advantages and disadvantages of the farm being close to an urban center. Marketing fish may be easier if the farm is close to the city. Also, input-markets should be accessible. On the other hand, farms close to the city may suffer from intruders, theft, and other problems.

ECONOMIC CONSIDERATION OF MANAGEMENT

Data Sources

Planning is concerned with future activities. In order to plan, one needs data. Governments and universities collect data through their research and extension centers and can provide good information. In lieu of an available source of data, the most important statistical source of information is the knowledge of seasoned, realistic specialists, both within and outside the government. Other means of obtaining information include questioning knowledgeable individuals in both written and personal interviews.

Regulations

Regulations and laws limit the type of fish species produced. Different states, communities, cultures, and nations have different regulations. There are regulations which govern the use of tilapia,

---

1. A method for determining land value is capitalizing annual returns from the land using the equation:

$$V = \sum_{n=1}^{\infty} \frac{\text{annual returns}}{(1 + i)^n} \quad \text{where } i = \text{prevailing capitalization rate}, \quad n = \text{year of returns in question}$$

carps, and other highly productive fishes in a production system. For example, regulations in Venezuela and Nepal influence the release of tilapia in ponds. Governments in several U.S. states do not allow the introduction of diploid grass carp and other species. There are also restrictions which limit the use of land, water, and the movement of fish. All regulations which affect aquacultural production must be considered before one decides to produce fish for commercial purposes.

Capital Requirements

Capital is one of the most limiting factors for individuals embarking on aquacultural projects. The level of capital required depends on the species, site, and level of production. Capital may be divided into long-term and short-term items. Long-term capital includes water rights, ponds, and reservoirs and more intermediate items such as storage bins, tractors, trucks, blowers, aerators, and fish-weighing scales. Short-term, or operating, capital is required if one hopes to produce fish for sale. Short-term capital also includes purchased inputs such as feed, fingerlings, and energy services. Operating capital also must cover costs of processing, marketing, and financing.

Skilled Personnel

Skilled personnel are essential if one hopes to undertake any large commercial venture. Aquaculture is a complex biological process and personnel with adequate management skills are needed to understand and manage the dynamic processes of fish production. In some cases, aquaculture will demand skills not required in other agricultural ventures.

Transferring an individual from an agricultural enterprise to aquaculture is not easy. While some aquacultural enterprises need only semi-skilled or unskilled labor, others require highly skilled individuals. To transfer individuals from one enterprise to the next requires training. Such training may form part of the management scheme.

Facilities and Layout

After the site has been selected, one must consider the physical layout of the fish farm. The layout will affect movement of capital, labor, fish, etc., and directly influence costs of operation. Ponds should be placed so that feeding, harvesting, and transportation are facilitated. One has to think of the optimal location relative to a processing plant, and accessibility to the plant. In the long-run, there might be plans for farm expansion. This should be given serious consideration early in the planning process. Other factors one should consider are electric power sources, waste treatment, processing plant capacity, and delivery systems.

Financial Considerations

After a site is selected and the plans are drawn, one must consider the financial aspects of the operation. Who will finance the operation, and how will it be financed? What type funds, such as equity capital, loans, and grants, are available? Detailed planning will help one in the acquisition of funds. One important factor is the return to capital. If the return to capital is lower than the prime interest rate, banks may not want to finance the project. Banks usually require detailed plans with total farm budgets, income statements, and cash flow statements. Additionally, financial ratios must be developed to show the bank this business can generate funds to pay for the capital used.

Marketing

Before investing in commercial aquaculture, the potential market for the product must be estimated. Demand may be either local, regional, export, or some combination of all three. The demand for the product, both immediate and future, will determine the type of facility and the fish species to be produced. Without adequate demand, any aquacultural project is bound to fail. Thus, the level of unmet demand must be estimated. Is it sufficient to guarantee a degree of success from the business venture?

Can the fish be sold in the local markets? If not, where will the fish be sold? Are there sufficient fish captured from the wild or imported from other sources to satisfy the market needs? How will these other sources of fish supply affect aquacultural production? Is there sufficient unmet demand to utilize both capture and produced fish? Or, alternatively, can the fish be produced at a price low enough to compete with other captured species? Adequate supplies of fresh fish may be available from local capture fisheries, or from a well-established trade network in the region. One should look at the kinds of fish available, the price compared to alternative sources of protein, seasonal demand, and variations in supply before deciding whether aquaculturally produced fish should be placed on the market.

Numerous special considerations are also important when evaluating the local market condition. For example, the competitive status of a new business venture could be affected by an active local cooperative. Cooperatives formed on the basis of either supply or product marketing needs, such as processing, may be the best alternative for emerging or limited resource operations. By working together to pool resources, better prices, market outlets, and processing and distribution systems may be available to co-op members.

Species Selection

The success of the fish species selected is influenced by the environment and region where the farm will be located. Whatever the aquaculture species considered for production, there are certain characteristics of organisms that lend themselves to commercial culture. Bardach (1976) suggested a few, but there are others including:

1. reproductive capacity, especially under the stress of captivity;
2. vigor of roe, spat, larvae, and fingerlings;
3. readily available foodstuffs at reasonable prices for each species considered;
4. relatively fast growth rate;

5. strong market demand for the species relative to competitive products, i.e., it can be sold at a reasonable price; and
6. desirability of taste and flesh texture of the species.

Farm Management Defined

Since management is so important, a definition should at least be attempted. A general definition may be readily developed, although identification of specific aspects may be difficult.

Management is the ability to organize resources in the most efficient and profitable manner. Managerial skills fall within the realms of both art and science. Art involves the application of thought processes to practical problems. Science is a learning process; it involves seeking information for problem solution.

Farm management concerns the making of farm business decisions that tend to maximize net farm income, consistent with the operator's or family's objectives. The objectives relate to business success but also include such items as health, education of children, increased standard of living, travel, and community activities. *Farm management is the mental effort directed toward the efficient organization and skillful operation of a farm business for the purpose of satisfying the goals of the farm family or operator*. Thus, the farmer is the key element in farm management. The very best resources may be of only limited value without skilled managers serving as farmers.

A *farm* is a parcel of real estate – land, buildings, plants, animals, fish, and other activities.

A *business* is a firm organized for the purpose of producing goods and/or services.

The *farm business* is an ongoing process which takes factors of production and combines them into a production process which will produce agricultural products.

The *farm family* is the complete labor/management unit responsible for combining the resources necessary for operating the farm. Although there may be alternative business forms such as partnerships, corporations, or cooperatives, it is still the individual farm family with its labor and skills that determines what resources will be available in what combinations for use in farm operations.

The *function of management* is to combine resources into a production process. Two basic phases of management necessary to the function are:

1. organizational – decisions which usually concern planning for long periods. The factors to consider in long-term planning of a fish farm or project have already been discussed.
2. operational – decisions regarding physical implementation of plans.

Leadership and initiative are essential elements in successful farm operations.

The *farm manager* makes those organizational and operational decisions for the farm business which may include but are not limited to:

1. what species of fish to produce;
2. how much of the selected species to produce;
3. what size and quantity to produce;
4. what mix of resources and technology to use;
5. when and where to sell, or buy; and
6. how to finance the operation?

FARM OPERATION

An operation is the implementation of planned activities to attain the goals of the farm family or entrepreneur. The *farm plan* is organized in the form of a farm budget in addition to the physical layout of the operation. The *farm budget* is a physical and financial plan for the operation of the farm. The total farm budget is prepared as an aid in organizing the entire farm. The main purpose is to compare the profitability of different enterprises which make up the farm organization.

FARM INCOME AND BUDGET ANALYSIS

What is Budget Analysis?

1. A budget is simply a plan to coordinate resource flows in and out of the business to achieve a given set of objectives.
2. Farm budgeting is concerned with organizing farm resources to maximize profits or, more often, family satisfaction.
3. Budget analysis involves several disciplines: farm management, economics, sociology, accounting, agriculture, biology, and other physical sciences. Budgeting is considered to be within the farm organization function of management in terms of efficiency and continuous profit.

Farm Income

1. "Income" is used interchangeably with revenue, receipts, sales, earnings, benefits, and sometimes profit; yet each has a different connotation. Profit is included in the terms for income although profit is quite different in many respects.
2. Definitions of farm income:
   a. Gross value of all goods produced on the farm—whether sold, consumed on the farm, or maintained as farm stock.
   b. Cash received from the sale of goods produced on the farm.
   c. Net receipts from the sale of products after expenses for inputs are deducted.
   d. Profit for an enterprise.
   e. Whole farm profit.

Analyzing the Farm Enterprises

1. Enterprises are defined as the different subdivisions or activities on a farm. Each is devoted to a particular kind of crop or animal, or a unique combination of the two.
2. Mixed enterprise systems refer to multiple use of the same resources to produce different farm outputs.
3. Enterprise studies help to explain the internal structure of the farm as a whole and to show the relative contribution of each enterprise to the whole organization. Often a whole farm busi-

ness may be showing a profit, but one enterprise might be doing badly.

4. Functions of enterprise studies:
   a. Examine the relation between individual enterprises and their relation to the whole farm.
   b. Assess the profitability of each enterprise relative to the resources used, even though the product may not be sold.
   c. Compare the relative efficiency of various enterprises on the farm.
   d. Compare the relative efficiency of enterprises on various farms similar in type, size, and farming conditions.
   e. Provide a basis for making rational decisions about the kind and size of enterprise for calculating production costs, and for setting prices of farm produce.

Measuring Enterprise Income

1. *Gross Output*
   a. Gross output is a *preliminary* measure of income. As such, it assesses the performance of an enterprise purely in terms of the benefits it yields without considering the costs to produce them.
   b. Gross output equals volume of final marketable product times average farm level price. Final production normally excludes intermediate products to avoid double counting, but intermediate products may be included in gross output when comparing gross output of individual enterprises. Farm level price—the point of first sale marking the dividing line between income derived purely from production and income from marketing the product—is used to calculate gross output.
   c. Since it is possible to produce more than one short-term crop within a year, a distinction should be made between gross output for a particular season and for the entire year.
   d. When brood or growing stocks are carried over from one production period to another, gross output may be defined more precisely as the *difference* between the closing value of stock (inventory) plus sales (including items consumed

on the farm or given away) *and* the beginning value of stocks (inventory) minus purchases. (See Table 7-1.)

2. *Net output*
   a. Net output is derived by subtracting the value of purchased inputs from the gross output of an enterprise. This measure is particularly significant for animal enterprises such as fish or livestock where purchased feed can be substituted for home-grown feed.
   b. Purchasing feed inputs is equivalent to increasing the area of land devoted to feed production. Hence, gross output may be much larger.
   c. Net output therefore provides a better measure for comparing the productive capacity of land devoted to fish production than gross output.
3. *Gross margin*
   a. Costs may be divided into fixed and variable expense items. Gross margin for an enterprise is found by subtracting vari-

TABLE 7-1. Estimated Gross Output of Tilapia Enterprise, 1 Hectare Pond Stocking 5000 per Hectare, Three Crops Yearly, 15% Mortality, Jamaica, 1988

Item	First crop	Second crop	Third crop	Total
Hectares	1.0	1.0	1.0	1.0
Yield per hectare (kilos)	890	890	890	2,670
Farm price per kilo	3.1	3.1	3.1	3.1
Gross output	2,759	2,759	2,759	8,277

Number and kind	Number and Kind	(*) Value	(=) Total
		dollars	unit of value
Closing inventory	25 brood fish	4.00	100
(plus) sales	2,670 kilos fish	3.10	8,277
(minus) beginning inventory	20 brood fish	4.00	(80)
(minus) purchases during year	5 brood fish	4.50	(22.50)
Gross output			8,274.50

able costs from gross output. For example, using data from Table 7.1 we see:

Gross output from sale of tilapia		$8277.
Fingerlings: 15,000 × $.0.05 each	750.	
Fertilizer: 1335 kgs. (2937 lbs.) × $0.60/kg. (2.2 lbs.)	801.	
Harvest cost: 2670 kgs. (5874 lbs.) × $0.10/kg. (2.2 lbs.)	267.	
Total variable cost		1818.
Gross margin (income above variable cost)		6459.

b. Since fixed costs are incurred regardless of production, they cannot be controlled in the short-run.

c. The gross margin is particularly useful in partial budgeting or linear programming.

d. Net Cash Income = Cash Farm Receipts − Cash Farm Expenses. Amount available from farm for family personal expenses including payment of debts, interest, and income taxes.

4. *Enterprise Profit*

a. Profit = Gross Output − Total Cost.

b. Gross margin excluded fixed costs; they are now included.

c. Labor costs provided by family members is considered hired.

d. Frequently, profit or net returns are calculated without subtracting land, operator's labor, or management. However, net profit must account for these items.

Mr. John's Farm

Let's look at the organization of a small farm business for Mr. John in Any County, USA. This example is based on an actual farm organization modified somewhat for teaching purposes. Mr. John has a 96-acre farm (approximately 38 hectares). The farm organization is shown in Table 7-2. He specializes in production of catfish, bass, and bream fingerlings. He adds some foodfish when price is favorable. Mr. John's farm also consists of a 12 acre reservoir, plus several smaller ponds and 49 acres of other lands, 4 acres of which are used as a homestead.

The land use pattern given in Table 7-2 shows that although only 35 acres of land are used in the production of fish, there is still some land which can be placed in fish production if Mr. John finds it profitable. He will know whether it is possible to increase production after conducting a budget analysis. As previously stated, a budget analysis will provide the entrepreneur with information on farm profitability. The first step in budget analysis is to take an inventory of Mr. John's farm (Table 7-2 and 7-3). In addition to cash on hand of $25,000, Mr. John owns land, equipment, and fish stocks as outlined below.

A look at the real estate inventory shows that the farm has a good pond system, hatchery, and buildings, together with real estate valued at $319,510 (see Table 7-4). Depreciation on buildings during the last year of operation was $1,170. The inventory of Mr. John's machinery and equipment is valued at $11,810 (see Table 7-5) after annual depreciation of $2,270 for the year. Inventory of feed and

TABLE 7-2. Land Use for Mr. John's Farm, Any County, U.S.A., 1991

Item	Acres
Catfish fingerlings	14.0
Bass fingerlings	8.0
Bream fingerlings	5.0
Catfish brood stock	2.0
Bass brood stock	1.0
Food fish	5.0
Reservoir	12.0
Other lands and homestead	49.0
	96.0

supplies showed that at the end of the year the value of feed and supplies was $1,075, a positive change of $671.00 (see Table 7-6).

Tables 7-7a, 7-7b, 7-7c, 7-7d, and 7-7e show Mr. John's output flows. The tables show the production, mortality, yield, purchases, and sale of Mr. John's business during the period January to December 31, 1991. Mr. John has fingerlings and brood stock which he may sell at the end of the year or later. Table 7-8 shows cash expenses for the year. Feed and labor are responsible for 44 percent of the cost. With this view of Mr. John's inventory, next examine his income statement for the year.

INCOME STATEMENT

The *income statement* is a summary of income and expenses over a given time period and is one type of financial statement needed for the management of the farm business. It is often called an *operating statement* or *profit and loss statement*, and its primary purpose is to compute profit for a given time period.

The period of time covered by an income statement is called the accounting period. For most farms, like that of Mr. John, annual income statements cover January 1 to December 31.

Income

Income may be in the form of cash or non-cash gains. Cash income includes payments received from selling commodities (fish) produced on the farm, and from other farm-related sources. Non-cash income may be accrued in the form of goods and services. One

TABLE 7-3. Direct Land Use by Enterprise, Mr. John's Farm, Any County, U.S.A., 1991

Item	Acres
Catfish (14 + 2)	16.0
Bass (8 + 1)	9.0
Bream	5.0
Foodfish	5.0
Total	35.0

TABLE 7-4. Inventory of Farm Real Estate for Mr. John's Farm, Any County, U.S.A., 1991

Item	Quantity	Beginning inventory	Depreciation	Ending value
	no.	dol.	dol.	dol.
Land[a]	96 acres	96,000	--	96,000
Pond system		210,000	--	210,000
Hatchery		6,455	645	5,810
Other buildings		3,000	400	2,600
Holding tanks	5	5,125	125	5,000
Well		100	--	100
		320,680	1,170	319,510

[a] No depreciation or appreciation in the land and the pond system during the year was assumed.

TABLE 7-5. Inventory of Machinery and Equipment of Mr. John's Farm, Any County, U.S.A., 1991

Item	Quantity	Beginning inventory	Depreciation	Ending value
	no.	dol.	dol.	dol.
Vehicles	4	5,000	1,000	4,000
Tractors	2	2,690	230	2,460
Four-wheel ATV		1,400	400	1,000
Well pump		640	240	400
Lift pump aerator		500	100	400
Hauling tanks	3	2,200	200	2,000
Mower & Blade		150	50	100
Nets 1,000		1,000	--	1,000
Spawning cans		500	50	450
		14,080	2,270	11,810

form of non-cash income on farms comes from farm-raised fish consumed by the farm household.

Expenses

Expenses may also be cash or non-cash in nature. Cash expenses include purchases of feed, fertilizer, seed, fingerlings, fuel, chemicals, etc. Non-cash expenses include depreciation on machinery,

TABLE 7-6. Inventory of Feed and Supplies for Mr. John's Farm, Any County, U.S.A., 1991

Item	Beginning Quantity	Value	Ending Quantity	Value
	no.	dol.	dol.	dol.
Feed (lbs)	1500	$210	3,000	$ 900
KMN_{04} (lbs)	50	94	40	75
Formalin (gal)	0	0	35	100
Liquid fertilizer (gal)	55	100	0	0
		404		1,075

Change in inventory +671.

TABLE 7-7a. Fish Inventory, Production for Mr. John's Farm, Any County, U.S.A., 1991

Item	Beginning	Value ($)	Production	Mortality
	no.	dol.	no.	no.
Catfish fingerlings	1,500,000	90,000	1,700,000	170,000
Bass fingerlings	0	0	18,000	1,500
Bream fingerlings	150,000	18,000	296,000	14,000
Catfish broodstock	600	13,200	--	0
Bass broodstock	100	2,450	--	0
Bream broodstock	550	2,750		10
Foodfish	0	0	20,000	1,000
		126,400		

TABLE 7-7b. Fish Purchased for Mr. John's Farm, Any County, U.S.A., 1991

Item	No. purchased	Unit price	Cost	Yield/acre
	no.	dol.	dol.	no.
Catfish fingerlings	20,000	.06	1,200	19,000
Bass fingerlings	0	--	--	2,201
Bream fingerlings	0	--	--	31,150
Catfish broodstock	20	5.70	104	--
Bass broodstock	20	7.00	140	
Bream broodstock	20	5.00	100	
			1,544	

equipment, buildings, and purchased broodstock. There is no money transaction for the latter. However, this decline in value due to depreciation is an expense to the business and must be included on the income statement.

An income statement does not include the cost of new capital assets such as tractors, buildings, pipes, pumps, and fences purchased during the year. Including their cost would distort profit for the year in which they were purchased, as these capital assets are used in future production. Instead, part of the purchase cost is included as a depreciation expense each year of its useful life. Also, applicable interest charges for this year may be reflected.

TABLE 7-7c. Sales and Replacement of Fish for Mr. John's Farm, Any County, U.S.A., 1991

Item	Replaced	No. sold	Unit Price	Value
			dol.	dol.
Catfish fingerlings	0	1,500,000	0.06	90,000
Bass fingerlings	0	16,500	0.50	8,250
Bream fingerlings	0	99,840	0.125	12,480
Catfish broodstock	100	100	5.20	520
Bass broodstock	10	10	7.00	70
Bream broodstock	0	0	--	--
Foodfish		19,000	.65	12,350
Total Value				123,670

TABLE 7-7d. Ending Value of Fish Inventory for Mr. John's Farm, December, 1991

Item	Ending number	Value
	no.	dol.
Catfish fingerlings	1,530,000	91,800
Bass fingerlings	0	0
Bream fingerlings	182,000	22,770
Catfish broodstock	620	13,640
Bass broodstock	120	2,940
Bream broodstock	560	2,800
		133,950

Cash Farm Income

Production activities of the farm business generate most of the cash farm income. Cash receipts from fish, brood stock, livestock, and crops are recorded in this section. Table 7-9 shows that cash farm income from the sale of fingerlings, broodstock, and foodfish was $123,670.

TABLE 7-7e. Total Assets, Mr. John's Farm, December, 1991

Asset	Value dol.
Real estate	319,510
Machinery & equipment	11,810
Feed and supply inventory	1,075
Fish inventory	133,950
Cash on hand	25,000
Total Assets	491,345

TABLE 7-8. Farm Cash Expenses for Mr. John's Farm for the Period January-December 1991

Item	Cost
	dol.
Labor	15,000
Repairs and maintenance	8,500
Gasoline, fuel, oil	11,000
Chemicals	1,200
Feed	15,000
Supplies	3,200
Broodstock	1,030
Fingerlings	1,200
Advertising	900
Accounting	310
Vehicle tags	150
Cash rent	500
Insurance	1,500
Utilities	4,400
Interest/principal payment	4,200
Taxes	400
Total	68,490

TABLE 7-9. Income Statement for Mr. John's Farm for Year Ending December 1991 (January 1 to December 31, 1991)

	Dollars	Dollars	Dollars
Cash farm income:			
Catfish fingerlings		90,000	
Bass fingerlings		8,250	
Bream fingerlings		12,480	
Cull broodstock		520	
Bass broodstock		70	
Foodfish		12,350	
Total cash income			123,670
Cash farm expenses (Table 7-8):			-68,490
Net cash farm income:			55,180
Non-cash adjustments:			
Depreciation:			
Machinery & equipment	2,270 (Table 7-5)		
Buildings	1,170 (Table 7-4)		
		-3,440	
Inventory change:			
Fish	7,550 (Table 7D-7A)		
Feed & supplies	+ 671 (Table 7-6)		
		+8,221	
Total non-cash adjustments			4,781
Net farm income (NFI)			59,961

Cash Farm Expenses

Two types of cash farm expenses are included in this section. Variable cash expenses include fertilizer, seed, feed, purchase of broodstock, labor, and supplies. Table 7-8 shows itemized cash farm expenses to be $68,490 for the year. The cost of newly purchased depreciable assets (such as machinery, buildings, and holding tanks) is not included as a cash expense, as these assets will be used in the business for more than one year.

It is also incorrect to include principal payments on loans as a cash farm expense. Interest paid on the loans is a business expense, but principal payments may be viewed as returning borrowed property and therefore not affecting profit. Rent and interest are similar in that each is a payment or expense required for the use of someone else's property.

Net Cash Farm Income

Net cash farm income is simply the difference between total cash income and total cash expenses. There has been no adjustment for non-cash items such as depreciation or inventory changes, and both adjustments can be large. In Table 7-9, the cash farm expenses of $68,490 is subtracted from the total cash farm income of $123,670 to give the Net Cash Farm Income of $55,180.

Non-Cash Adjustments to Income

The final section of the income statement includes all non-cash adjustments to income. Depreciation is a non-cash expense, and this adjustment will always be a negative entry reducing the year's profit. In Table 7-9, depreciation for machinery, equipment and buildings is (negative) $3,440.

The second major non-cash adjustment to income is the net change in inventory value. The assets grown or raised on the farm, those purchased for resale, and supplies such as feed and fertilizer are included. Capital assets such as buildings and machinery are not included as they have been accounted for by the depreciation expense. In Table 7-9, the change in inventory for fish, feed, and supplies amounts to positive $8,221.

The non-cash adjustment to income is $4,781. This income statement does not show a value for products consumed on the farm. Any farm products consumed would *increase* the value of non-cash adjustments.

Net Farm Income

Net farm income (NFI) is derived by adjusting net cash farm income for total depreciation, net inventory changes, and value of fish products consumed in the home (zero in this case). This is the true measure of profit for the accounting period, as net cash farm income does not include the listed adjustments which can be quite large. Net farm income represents the return to the owner for personal labor, management, and equity capital used on the farm. An NFI of $59,961 is calculated from Table 7-9 by adding $4,781 (non-cash adjustments) to $55,180 (the net cash farm income).

Analysis of Net Farm Income

Profitability is also found by calculating NFI. Profitability is a measure of the relative volume of the business or the value of the resources used to produce the profit. A business may show a positive profit, but have a low profitability rating if profit is small relative to the size of the business. To determine the profitability of Mr. John's operation, ratios based on size of operation and investment must be calculated.

Return to Capital

The return to capital, or the return on investment, is a measure of profitability obtained by dividing the return to total capital by total farm assets. It is expressed as a percentage to allow easy comparison with returns from other investments:

$$\text{Rate of return to capital} = \frac{\text{return to total capital}}{\text{total farm assets}}$$

Return to capital is the dollar return to both debt and equity capital, so net farm income must be adjusted. The interest on debt capital was deducted as an expense in calculating net cash income. This interest must be added back to net farm income before the return to capital is computed. From Table 7-8, the interest paid of $4,200 is added to NFI to obtain a total adjusted NFI of $64,161. NFI includes the return to the owner's labor and management as well as the return to all capital. Therefore, a return to owner's labor and management must be subtracted from the adjusted NFI to find the actual return to capital in dollars. Assuming the opportunity cost of Mr. John's labor is $10,000 and management is $2,000, the return to capital would be:

Adjusted net farm income	$64,161
Less opportunity cost of labor	−10,000
Less opportunity cost of management	−2,000
Equals return to capital	$52,161

For Mr. John, the percent return on capital is:

$$\frac{52{,}161}{491{,}345} \times 100 = 10.6\%$$

Return to Labor and Management

The return to labor and management is expressed in dollars. It is another measure of profitability and is the portion of NFI which is used for payment of the owner's personal labor and management.

If the opportunity return for Mr. John's capital is 8 percent, the opportunity cost of capital is \$491,345 × .08 = \$39,307. Mr. John's labor and management earning is \$64,161 − \$39,307 or \$24,854 net of opportunity cost:

Adjusted net farm income	\$64,161
Less opportunity cost on total capital	−39,307
Equals return to labor and management	\$24,854

Return to Labor

The return to labor and management can be used to compute a return to labor alone. Since the opportunity cost of total capital has already been subtracted from adjusted net farm income, the only remaining step is to subtract the opportunity cost of management:

Return to labor and management	\$24,854
Less opportunity cost of management	−2,000
Equals return to labor	\$22,854

Return to Management

Management is often considered the residual claimant to NFI. Its opportunity cost is a simple estimate. The return to management can be found by subtracting the opportunity cost of labor from the return to labor and management:

Return to labor and management	\$24,854
Less opportunity cost of labor	−10,000
Equals return to management	\$14,854

Return to Equity

The return to capital was a return to both debt and equity capital. Mr. John may be more interested in the return to personal equity capital invested in the business than the return to capital.

The calculation of return to equity begins directly with net farm income, as no adjustment is needed for any interest expense. Interest is the payment or return to borrowed capital, which must be properly deducted as an expense before the return to equity is computed. However, the opportunity cost of labor and management must be subtracted to find the dollar return to equity.

The dollar return to equity would be:

NFI	\$59,961
Less opportunity cost of labor	−10,000
Less opportunity cost of management	−2,000
Equals return to equity	\$47,960

Equity is the value in total assets which exceeds debts owed. If Mr. John had outstanding debts of \$160,470, his equity would be \$330,875.

$$\text{Rate of Return to Equity} = \frac{\text{return to equity}}{\text{net worth of owners's equity}} \times 100$$

$$= \frac{47{,}960}{330{,}875} \times 100 = 14.5\%$$

$$\text{The debt to asset ratio} = \frac{\text{total debts}}{\text{total assets}}$$

$$= \frac{160{,}470}{491{,}345} = 32.7\%$$

which appears about normal for this type of farm business.

DEPRECIATION

Most capital assets decrease in value with use and the passage of time. Machinery, buildings, and other equipment are among the farm assets subject to depreciation, and the farm manager must explicitly consider this phenomenon since it has a direct effect upon both efficiency and costs. Normally, aquaculture ponds are not considered as depreciable assets since they are part of the real estate. However, determination of whether an item is depreciable depends on the government under which the business is organized, since much of the issue over depreciation is related to income and land tax laws. Even so, there are certain predictable rates for asset loss of economic life. Thus, the prudent manager will account for depreciation in the business regardless of whether there are tax laws affecting the outcome.

Depreciation results from three principal causes: (1) Physical deterioration – as a machine or structure is used or subjected to the passage of time, elements wear out; (2) technological obsolescence – new equipment or processes may be developed that will perform the same job at less cost, produce a higher quality product, or be a safer machine or process; or (3) economic obsolescence – the demand for the product produced by the machinery in question may no longer exist.

There are three principal ways of depreciating assets: the straight-line, declining balance, and the sum-of-the-year's digits methods. The straight-line method of calculating depreciation is the most widely used and the easiest to use. In the United States, an accelerated cost recovery system is allowed for income tax purposes. However, that is not covered here since it is specific to only one country.

Straight-Line Method

Annual depreciation is computed from the equation:

$$\text{Annual depreciation} = \frac{\text{Cost} - \text{Salvage Value}}{\text{Useful Life}}$$

and is the same for each year. Straight-line depreciation can be computed from an alternate method, using the equation:

$$\text{Annual depreciation} = (\text{Cost} - \text{Salvage Value}) \times R$$

where R is the annual percentage depreciation rate found by dividing 100 percent by the useful life (100 percent/useful life).

Example: Assume the purchase of a tractor for $10,000 which is assigned a $2,000 salvage value and a 10-year useful life. The annual depreciation using the first equation would be:

$$\frac{10{,}000 - 2{,}000}{10} = \$800$$

Using the second equation, the percentage rate would be 100 percent divided by 10, or 10 percent, and the annual depreciation is:

$$(10{,}000 - 2{,}000) \times 0.10 = \$800$$

Declining Balance Method

Several types of declining balance depreciation are possible, but the most common is simple declining balance. The "simple" comes from using a depreciation rate which is similar to the straight-line rate, except a fixed rate is used rather than one based on useful life. A double rate is sometimes used on some types of property when calculating depreciation for income tax purposes. Annual depreciation may be computed from the equation:

$$\text{Annual depreciation} = (\text{Book value at beginning of yr.}) \times R$$

where R is equal to the straight-line percentage rate. The percentage rate remains constant each year, but it is multiplied by the book (residual) value, which declines each year by an amount equal to the previous year's depreciation. That residual value is used each succeeding year to calculate depreciation.

Notice also that the percentage rate is multiplied by each year's book value and not cost minus salvage value as with the straight-line method.

Using the previous example, the declining balance rate would be 10 percent, and the annual depreciation would be computed in the following manner (decimal numbers are rounded):

Year 1:	$10,000	× 10% =	$1,000
Year 2:	9,000	× 10% =	900
Year 3:	8,100	× 10% =	810
Year 7:	5,314	× 10% =	531
Year 8:	4,783	× 10% =	478
Year 10:	3,874	× 10% =	387
Year 15:	2,288	× 10% =	229
Year 16:	2,059	× 10% =	206

(But this amount (206) would reduce the book value below the 2,000 salvage, so only $59 of depreciation may be taken).

Year 17 - 20: No remaining depreciation

If double declining balance procedures were used, the rate, R, usually is two times the straight-line rate. This example with a double rate would allow depreciation to be completed in eight years. This situation is not unusual, as double declining balance will often result in the total allowable depreciation being taken before the end of the useful life, and depreciation must stop when the book value equals salvage value. With zero salvage value, it is necessary to switch to straight-line depreciation at some point to get all the allowable depreciation or to take all remaining depreciation in the last year.

Sum-of-the-Year's Digits

The annual depreciation using the sum-of-the-year's (SOYD) digits method is computed from the equation:

$$\text{Annual depreciation} = (\text{Cost} - \text{Salvage Value}) \times \frac{\text{RL}}{\text{SOYD}}$$

where:

RL = remaining years of useful life of the beginning of the year for which depreciation is being computed.

SOYD = sum of all the numbers from 1 through the estimated useful life.

For example, for a 5-year useful life, SOYD would be $1+2+3+4+5=15$ and would be 55 for a 10-year useful life (add numbers 1-10). For example:

$$\text{Year 1:} \quad (10{,}000 - 2{,}000) \times \frac{10}{55} = 1{,}454.55$$

$$\text{Year 2:} \quad (10{,}000 - 2{,}000) \times \frac{9}{55} = 1{,}309.09$$

$$\text{Year 3:} \quad (10{,}000 - 2{,}000) \times \frac{8}{55} = 1{,}163.64$$

$$\text{Year 10:} \quad (10{,}000 - 2{,}000) \times \frac{1}{55} = 145.45$$

A quick way to find the sum-of-the-year's digits (SOYD) is from the equation:

$$\frac{(n)(n+1)}{2}$$

where n is the useful life. In this case it is:

$$\frac{10(10+1)}{2} = 55$$

Summary of the Valuation Methods

The valuation method or methods to use will depend not only on the type of property, but also on the final use of the inventory values. Consistency in the valuation method selected is important.

Useful life is the expected number of years the item will be used in the business. It may be the age at which the item will be completely worn out.

Salvage value or *terminal value* is the value of the item at the end of its assigned useful life. Salvage value may be zero if the item will be owned until completely worn out, and will have no junk or scrap value at that time.

PARTIAL BUDGETING

In organizing the farm, the total farm budget was considered, which required a lot of time and hard work. Sometimes only a minor change is required on the farm. If so, a partial budget may be developed to look at the profitability of that change. Partial budgeting is used to evaluate expected changes in costs and returns by introducing some change into the production system. The change may occur by *intensification* of input, by changing the *technology*, or by increasing *efficiency*. Partial budgeting relates to marginal concepts in that only the changes are evaluated.

Partial budgeting is always performed ex-ante, or before the change is made. Data for evaluating changes may come from experimental results or results obtained from extension demonstrations. Other information might be implied from existing production practices. Another source of information would be similar production activities in a different geographic area. Technology does not transfer intact from area to area; however, estimates of the effect can be made from similar work elsewhere.

Since a partial budget is used to calculate the expected change in profit for a proposed change in farm business, it is useful to think of partial budgeting as a type of marginal analysis, i.e., adapted to analyzing relatively small changes in the whole farm plan. A partial budget contains only those income and expense items which will change if the proposed modification in the farm plan is implemented. Only the changes in income and expenses are included, not the total values. The final result is an estimate of the gain or loss in profit. If the firm has an inventory of unused resources—for example: labor, land, capital, organic fertilizer, etc.—partial budgeting is especially appropriate in evaluating resource use.

Experimental data indicate that yield is linear with increased stocking rate so long as water quality is maintained. Thus, change must be evaluated from four aspects: (1) any increase in costs, (2) any decrease in returns, (3) any decrease in costs, and (4) any increase in returns. These aspects cover changes in returns. If returns are greater than costs, then the farmer may choose to make the change in production. Suppose the farmer wishes to evaluate the feasibility of increasing the tilapia stocking rate from 6,625 to 8,000 per hectare (2.5 acres). Table 7-10 shows that by increasing the stocking rate by 1,375 fingerlings per hectare, the net return will increase by $223.35. Alternatively, if the same stocking rate was maintained, but the feed quality was reduced to 30 percent protein, the net return due to the change would be only $32.49. Such a small change in income would make the protein switch questionable.

A widely used partial budgeting format is illustrated in Figure 7-1. It contains four basic headings to organize the information relating to the following four questions:

1. What new additional costs will be incurred?
2. What current income will be lost?
3. What new or additional income will be received?
4. What current costs will be reduced or eliminated?

TABLE 7-10. Partial Budget of a Change in Stocking Rate of Tilapia from 6,625 per Hectare to 8,000 per Hectare

			dollars
1.	Increase in cost		
	Fingerlings	1,375 @$.12 each	165.00
	Feed	1.18 M.T.@343/MT	404.74
	Fuel, oil, etc.	12 units @$1.00	12.00
	Depreciation		
	Tractor		4.80
	Feed storage		4.80
	Interest on operating loan	591 @ 14%/yr.(4mo.)	27.58
	TOTAL		618.92
2.	Reduction in return		0
3.	Increase in return	589 kg @ 1.43/kg	842.27
4.	Reduction in cost		0
5.	Net return		223.35

Alternatively, consider maintaining the same stocking rate and reducing the feed quality to 30% protein level.

1.	Increase in cost		0
2.	Reduction in return	114 kg @$1.43/kg	163.02
3.	Increase in return		0
4.	Reduction in cost	Feed 5.7 M.T. @ $34.30/M.T.	195.51
5.	Net return		$32.49

FIGURE 7-1. Partial Budgeting Format

Proposed change____

____

Additional cost ($)	______	Additional income ($)	______
Reduced income ($)	______	Reduced costs ($)	______
A. Total annual additional costs and reduced income	$ - ______	B. Total annual additional income and reduced costs	$ ______
			$ - ______
		Net change in profit (B-A)	$ ______

Additional costs are the additional expense associated with the proposed change.

Reduced income is a listing of all receipts that would no longer be obtainable under the alternative plan.

Additional income is an estimate of additional receipts that will occur from the proposed change.

Reduced cost is a listing of the inputs and their values which will no longer be incurred if the change is made.

From this, the net change in costs and receipts are calculated. The format shown in Figure 7-1 will be used to study a proposed change to Mr. John's farm in Any County, U.S.A.

Mr. John is thinking of putting 5.0 acres of water into catfish. He does not want to make any major farm readjustments. He simply wants to change the 5.0 acres which are already used for production of bream fingerlings to catfish. Mr. John thinks that it is wise to measure the effects of such a change on his income before he begins. He makes certain assumptions about the production process of catfish:

1. the catfish will be produced using recommended management practices of the State Agricultural Experiment Station;
2. the stocking rate will be 3,500/acre instead of the previous 4,000/acre used;
3. an anticipated feed conversion rate of 1.7 pounds for catfish;
4. 200 days in the growing season; and
5. 6 percent death loss.

Mr. John assumed that the fixed costs allocated for bream production will be approximately the same for the 5.0 acres of catfish. Therefore, he was considering only the changes in variable costs associated with the change.

The costs and returns associated with the 5.0 acres of bream are shown in Table 7-11. The total annual cost and reduced income for the change is $21,516.69. The additional cost was taken from the production of 5.0 acres of catfish, using recommended practices. Income would be reduced by the elimination of last year's sale of bream fingerlings plus the sale of broodstock bream presently used on the farm. The additional income was derived from the sale of catfish from the 5.0-acre pond. Cost would be reduced by the expenses not used in producing bream. The difference between A and B is a negative $8,687.94. This means that the replacement of the 5.0 acres of bream fingerlings with 5.0 acres of catfish will reduce NFI, and is therefore not a wise decision.

Mr. John did not have to develop a total farm budget to determine the effects of the anticipated change on his NFI. The partial budget was sufficient for the purpose.

THE ENTERPRISE BUDGET

The enterprise budget is a process of estimating costs and returns for a particular activity. It is a statement of what is generally expected from using particular production practices to produce a specified amount of product. An estimate of expected revenues and expenses is needed. The enterprise budget contains all levels of output and their projected price, as well as the level of all inputs and their costs for that enterprise. Enterprise budgets generally include in-

TABLE 7-11. Partial Budget for an Additional 5 Acres of Catfish on Mr. John's Farm

Additional cost $		Additional income	
Fingerlings (4") 17,500 x .05	= $875.00	Sale of food fish 16,450 x 65	= 10,692.00
Floating feed 14.3 tons	= 4,098.06		
Chemicals	= 350.00		
Harvest labor	= 64.00	Reduced cost	
Tractor (fuel, oil, lube)	= 68.00	Bream broodstock	= 170.00
Electricity	= 378.00	Feed for broodstock	= 180.00
Machinery & equip. (repair)	= 156.00	Feed cost .3t x 300 x 5	= 450.00
Interest on operating cap.	= 246.95	Fertilizer 25 gP x 1.80 x 5	= 225.00
		Chemicals $70 per ac	= 350.00
Total	$ 6,236.69	Harvest labor	= 64.00
		Tractor (fuel, oil, lube	= 46.00
Reduced income		Electricity	= 260.00
Sale of Bream fingerlings =	$12,480.00	Machinery & Equip. oper.	= 180.00
Sale of Bream broodstock =	2,800.00	Interest on op. capital	= 211.75
			$2,136.75
Total reduced income =	$15,280.00		

A. Total annual additional cost and reduced income = $21,516.01

B. Total annual additional income and reduced cost = $12,828.25

Net change in profit (B-A) = ($-8,687.76)

come, variable costs, and fixed costs. An enterprise budget for channel catfish in Alabama will be used to show the various sections of this statement, illustrated by Table 7-12.

The budget is for catfish produced from 5.0 acres of existing ponds on the assumption that catfish is not the main farm activity. This allows for sharing certain general machinery and equipment items which may be available on a typical farm. The pond is further assumed to be a hillside pond, constructed by building a dam across a narrow valley.

The first step in developing an enterprise budget is to estimate the total production and the expected output price. The total value of the catfish sold is called the *gross receipts*. The fish average 1.0 pound each when sold. It is estimated that 6.0 percent of the fingerlings stocked die or are not recovered at harvest.

The second step is estimating the *variable costs*. Variable costs are the cash expenses directly related to production. These costs differ for the various size ponds and stocking rates. Variable expenses include cost of fingerlings, feed, chemicals, harvest labor, electricity, equipment repair, and interest on operating capital.

The third step is to calculate the *income above variable costs*. This indicates the income above cash costs.

The fourth item is the *fixed costs*. Fixed costs are incurred regardless of whether production occurs. Certain items must be purchased which last longer than one production period. Only expenses related to equipment and machinery are considered fixed costs in these budgets because it was assumed that ponds were already constructed.

The fifth step is to calculate total costs, which is the sum of total variable and total fixed costs. The sixth item is the returns to land, labor, capital, and management.

Assumptions for the catfish production budget shown in Table 7-12 were:

1. There are existing ponds.
2. Fingerlings (4 inches) were stocked in the spring (March).
3. There is a willing buyer for the fish.
4. All systems had a 6.0 percent loss (death plus unharvested fish).

5. Electric aerators were used in all pond sizes. Operating time is calculated to be six hours per night for 180 days, starting in May. Emergency aeration (P.T.O. driven) is also required with duration varying with pond size and stocking rate.
6. The fish production enterprise is only one enterprise on a commercial farm. Equipment shared between enterprises was charged to the catfish enterprise on a percentage-used basis.
7. The operator supplies all routine management and labor. Additional labor is hired to complete the harvest where indicated.
8. The fish are harvested by the buyer in November.
9. 3,500 fish stocked per acre with 4-inch fingerlings, 20 pounds/1000 beginning weight.
10. Feed ration – 1.7 pounds of feed/pound of gain over 200 days in a growing season.
11. Ending weight of 1 pound.

BREAK-EVEN ANALYSIS

Knowing the break-even cost of production and how costs are allocated between fixed and operating costs is highly useful. Unless a producer is aware of production costs, it is difficult to accurately calculate prices. The nature of break-even analysis is depicted in Figure 7-2. The volume of output is measured on the horizontal axis, and revenue and cost are shown on the vertical axis. Since fixed costs are constant regardless of the output produced, they are indicated by the horizontal line. Variable costs at each output level are measured by the distance between total cost curves and the constant fixed costs. The total revenue curve indicates the price times the quantity demanded for the firm's product, and the profit is shown by the distance between the total revenue curve and the total cost curve.

A linear break-even chart as illustrated in Figure 7-3 is usually more common in business. It is easy to use and facilitates quick calculations. In Figure 7-3, the break-even quantity and break-even price are easily identified. The profit and loss areas are shown as the shaded portion.

TABLE 7-12. 5-Acre Catfish Budget (Existing Ponds): Estimated Costs and Returns; Using Recommended Management Practices: Alabama, 1990

Item	Weight each	Unit	Quantity	Price or cost/unit	Value or cost
1. Gross receipts					
Catfish	1.00	lbs	16450.00	.65	10692.50
2. Variable cost					
Fingerlings (4")		each	17500.00	.05	875.00
Floating feed (32%)		ton	14.13	290.00	4098.06
Chemicals		apl/ac	1.00	70.00	350.00
Harvest labor		hr	16.00	4.00	64.00
Tractor (fuel, oil, lube)		hr	34.00	2.00	68.00
Electricity		kwh	5400.00	.07	378.00
Mach. & equip. (repair)		dol			156.68
Interest on operating cap.		dol	2244.95	.11	246.95
Total variable cost					6236.69
3. Income above variable cost					4455.81
4. Fixed cost					
General overhead		acre	5.00	5.00	25.00
Int. on bldg. and equipment		dol 4345.13	.115	499.69	
Depr. on bldg. and equip. dol.					748.21
Other f.c. on bldg. & equip.		dol		62.59	
Total fixed costs					1335.49
5. Total of all specified expenses					7572.18
6. Net returns above all specified expenses					3120.32

Net returns per acre:	Above specified variable expenses	891.16
	Above specified total expenses	624.06
Breakeven price (per cwt. sold):		
	To cover specified variable expenses	37.91
	To cover specified total expenses	46.03

FIGURE 7-2. Break-Even Quantity, Price, Profit, and Loss for Catfish, Assuming a Curvilinear Cost Function

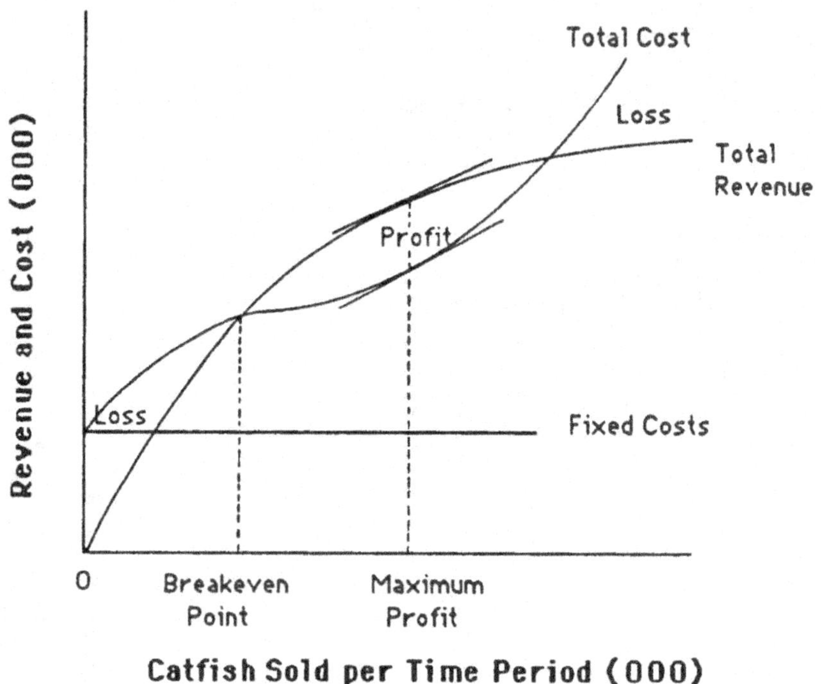

Calculation of Break-Even Cost

Break-even cost is calculated using the formula:

$$\text{Break-even cost} = \frac{\text{cost per unit of production}}{\text{yield per unit of production}}$$

The catfish enterprise budget in Table 7-12 shows that variable or operating costs for a 5.0-acre pond were $6,236 and the fixed cost was $1,335.49. The total cost for the 5.0-acre pond amounted to $7,572.18. The yield for the 5.0-acre pond was 16,450 pounds. The break-even cost per pound is calculated thus:

FIGURE 7-3. Break-Even Quantity, Price, Profit, and Loss for Catfish, Assuming a Linear Cost Function

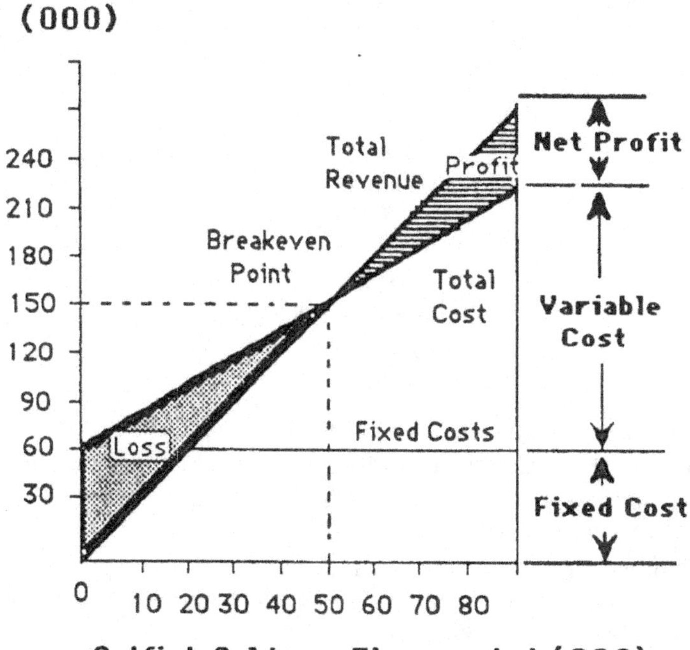

$$\text{Break-even cost per lb.} = \frac{\$7{,}572.18}{16{,}450} = 46.03 \text{ cents}$$

This means all fish produced must be sold for a minimum 46.03 cents per pound to cover all costs. The amount attributed to variable cost may be calculated in this way:

$$\text{Break-even price to cover variable expense} = \frac{\text{Operating cost}}{\text{Yield}}$$

$$= \frac{6{,}236}{16{,}450} = 37.91$$

Thus, to cover only variable costs, all the fish produced on 5.0 acres would have to be sold for 37.91 cents per pound. That is, 37.91 cents per pound are attributed to the variable costs portion of the total 46.03 cents. The break-even price to cover fixed expenses is calculated in the same way.

$$\text{Break-even price to cover fixed expense} = \frac{\text{Fixed Cost}}{\text{Yield}}$$

$$= \frac{1{,}335}{16{,}450} = 8.12 \text{ cents}$$

This means that 8.12 cents are attributed to fixed costs.

At some time in the operation, one may wonder whether it is profitable to produce in a given year due to exceptionally high input cost or low produce price.

Previously it was noted that it generally pays to produce in the short-run, as long as operating costs can be covered. Any income over operating expenses may then be applied toward fixed expenses, since they will continue regardless of whether production continues. Sometimes, therefore, it is important to know what levels of production will cover costs. From the catfish enterprise budget shown in Table 7-12 calculate the following ratios:

$$\text{Break-even production level in pounds to pay operating expenses} = \frac{\textit{Operating expenses}}{\textit{Expected Selling Price}} = \frac{6{,}236}{.65} = 9{,}594 \text{ lbs}$$

$$\text{Break-even production level in pounds to pay fixed expenses} = \frac{\textit{Fixed expenses}}{\textit{Expected Selling Price}} = \frac{1{,}335}{.65} = 2{,}054 \text{ lbs}$$

$$\text{Break-even production level in pounds to pay all expenses} = \frac{\textit{Total expenses}}{\textit{Expected Selling Price}} = \frac{7{,}572}{.65} = 11{,}649 \text{ lbs}$$

THE BALANCE SHEET

The farmer or farm manager should be interested in knowing the financial condition of the business at any point in time. The balance sheet, or net worth statement, will at any point in time give some indication as to the financial state of the business. The manager and credit source will want to know whether the business is solvent. The *balance sheet* represents a snapshot of the firm's position on a given date, whereas the income statement developed earlier is based on a few concepts showing what occurred between two points in time. The balance sheet is a systematic organization of everything "owned" and "owed" by a business or individual at a given point in time. Thus, it includes a listing of assets and liabilities, concluding with an estimate of net worth or owner's equity which is the difference between total assets and total liabilities:

Net Worth = Assets − Liabilities,

or said another way:

Assets = Liabilities + Net Worth.

Assets

Assets are properties owned by or owed to the business enterprise. Assets are usually divided into three categories: current, intermediate, and fixed.

Current assets are those which may be readily sold for cash. Goods which have already been produced (such as fingerlings, fish, and feed) may be sold quickly for cash without disrupting ongoing production activities. These assets are said to be *liquid*. Cash, checking accounts, savings, and accounts receivable in a time period of less than one year are said to be current.

Intermediate assets are those that may be sold for cash in a time period of more than one year but less than 10 years, and are less liquid than current assets. Machinery, tractors, aerators, breeding stock and feeders are considered intermediate assets. Most intermediate assets are characterized by being depreciable.

Fixed assets are those which require a long period for sale (such as land, buildings, and more permanent fixtures) which, if sold, would affect the nature of the business.

Equity refers to the unencumbered rights or claims by owners to the property and assets regardless of time period. Equity is another term for net worth.

Liabilities

Liabilities are obligations or debts owed to someone else. They are an outsider's claim against one or more business assets. Liabilities are also divided into current, intermediate, and fixed.

Current liabilities are those financial obligations which will become due and payable within 1 year. Accounts (credit lines), accrued expenses (such as taxes, rent/lease payments, and interest on notes payable in less than a year), short-term loans, as well as interest and principal on a loan payable during the year are considered current liabilities.

Intermediate liabilities are those loans where repayment is extended over at least 1 year and up to as long as 10 years. Most of these will be loans made for purchasing machinery, broodstock, or other intermediate assets.

Long-term liabilities are debts which occur for the purchase of long-term assets. They will be in the form of a farm mortgage loan or land purchase contract. The repayment will generally be from 10 to 40 years.

Net Worth

Net worth represents the amount of money left for the owner of the business should the assets be sold and all liabilities paid on the date of the balance sheet. Net worth is the owner's current investment or equity in the business.

In preparing a balance sheet, the assets are arranged in order of decreasing liquidity; that is, assets listed first will be converted to cash sooner than those toward the bottom of the columns. The top group of assets consists of cash marketable securities, accounts receivable, and inventories which are expected to be converted into cash within 1 year. Then the intermediate assets are listed and finally the fixed assets. The liabilities are then listed in the same order. Refer to the balance sheet for Mr. John in Table 7-13.

Mr. John's business organization is examined on December 31, 1990. The balance sheet shows that total assets are larger than total liabilities. This does not tell much unless some financial ratios are calculated.

TABLE 7-13. Balance Sheet for Mr. John, December 31, 1991

Assets			
Current assets	........dollars........		
Bank balance	25,000		
Fish fingerlings	114,570		
Feed and supplies	1,075		
Total current assets		140,645	
Intermediate assets			
Machinery & equipment	11,810		
Broodstock	19,380		
Total intermediate assets		31,190	
Fixed assets			
Land	96,000		
Pond system	210,000		
Hatchery & building & well	13,510		
Total fixed assets		319,510	
Total assets			491,345
Current liabilities			
Operating loan	52,500		
Interest on operating loan	4,200		
Principal payment on mortgage	2,500		
Interest on mortgage	3,000		
		62,200	
Intermediate liabilities			
Payment on equipment	2,270		
Long-term liabilities			
Mortgage	96,000		
Total liabilities		160,470	
Net worth			330,875

Financial Ratios

Financial ratios are used to analyze the balance sheet. Financial ratio analysis is an exercise undertaken to understand the liquidity, credit worthiness, solvency, efficiency, and profitability of the business. Trends over time and comparisons with other firms will give some idea of how well the business is doing.

To begin, the solvency of the farm business is considered. Solvency (liquidity) is used to evaluate the current position of the business relative to all assets and debts. Solvency ratios may be computed directly from the balance sheet. Following solvency, concern should be focused on business liquidity. Ratios to evaluate solvency and liquidity are presented below.

Net Capital Ratio

The net capital ratio is a measure of the overall financial strength and solvency (liquidity) of the farm. The ratio is calculated as:

$$\text{Net capital ratio} = \frac{\text{total assets}}{\text{total liabilities}}$$

From Mr. John's farm the net capital ratio is:

$$= \frac{491{,}345}{160{,}470} = 3.06$$

The net capital ratio reflects the likelihood that the sale of all assets would produce sufficient cash to cover all outstanding debts. Thus, a value for this ratio greater than 1.0 indicates that business liquidation would provide enough cash to pay all liabilities.

Debt-to-Equity (Leverage) Ratio

An additional measure of solvency commonly used is the leverage ratio. This ratio helps to evaluate the debt relative to owner's equity. Thus, it is useful when financing investments. As one proceeds with debt payments, this ratio should decrease. The ratio is calculated as follows:

$$\text{Debt: Equity Ratio} = \frac{\text{total debt}}{\text{owner's equity}}$$

$$= \frac{160{,}470}{330{,}875} = .48$$

The smaller the value, the larger net worth or equity is relative to total liabilities. Lenders typically do not set rigid guidelines on minimum or maximum ratios. However, it should be kept in mind that a value greater than 1.0 means that the lenders have more invested in the business than the farmer.

In addition to solvency, there is concern that the business is able to meet current obligations without interrupting the ongoing operation. Having insufficient *liquidity* to pay harvest costs would be disastrous for an aquacultural operation at harvest time. Thus, it is useful to evaluate the cash position. This is done through the current ratio and the quick ratio.

Current Ratio

This is a measure of the extent to which current assets, if liquidated, would cover current outstanding debts. It is calculated as follows:

$$\text{Current ratio} = \frac{\text{current assets}}{\text{current liabilities}}$$

$$= \frac{140{,}645}{62{,}200} = 2.26$$

For the banker, the current ratio is intended to show the margin the enterprise or business has for its current assets to shrink in value before there is difficulty in meeting current obligations. If the current ratio is low, the enterprise may be existing on a day-to-day basis, which means it has to follow uneconomic practices that could affect the safety of any proposed new project. A rule of thumb sometimes applied by traditional bankers is that the current ratio should be approximately 2:1. Mr. John's farm is able to meet its current liabilities.

Quick Ratio

Almost identical to the current ratio, the quick ratio goes one step further to exclude inventories, supplies, and cash invested in fish. Thus, the assets listed under the current ratio are reduced by the value of these items. The purpose for this adjustment is to account for businesses such as fish farms which carry large inventories which may not be marketable immediately, but require harvest and liquidation at the end of the growing cycle. Application of a quick ratio is not essential for most aquacultural operations since most lenders are willing to wait until harvest periods. However, it is a useful bit of information, especially when fish are grown on a continuous basis rather than year-to-year.

$$\text{Quick ratio} = \frac{\text{current assets} - \text{inventories, supplies, and cash in fish stocks}}{\text{current liabilities}}$$

Intermediate or Working Capital Ratio

This ratio reflects the ability of the business to meet debts over a 1-10 year period. It is calculated as:

$$\text{Intermediate ratio} = \frac{\text{current} + \text{intermediate assets}}{\text{current} + \text{intermediate liabilities}}$$

$$= \frac{171{,}835}{64{,}470} = 2.26$$

In some texts, the intermediate ratio is calculated by dividing only intermediate assets by intermediate liabilities. The method used to calculate the intermediate ratio depends on the intended use of the information. If the need is for only an assessment of the status of intermediate assets, then the ratio of intermediate assets and liabilities is appropriate. However, if the need is for an evaluation of total working capital, then the combined current and intermediate assets and liabilities is preferred.

Equity Value Ratio

This ratio measures leverage that the owner may have in borrowing money:

$$\text{Equity value ratio} = \frac{\text{owner's equity}}{\text{value of assets}}$$

$$= \frac{330{,}875}{491{,}345} = .673$$

Fixed Ratio

The fixed ratio allows examination of the relative load on the business by fixed long-term debt such as a mortgage. Mr. John has a $96,000 mortgage. Does he have sufficient value in real estate to cover that debt? The answer is yes. His mortgage load is 30 percent of the real estate value.

$$\text{Fixed ratio} = \frac{\text{long-term fixed debt (mortgage)}}{\text{security for that debt (real estate value)}}$$

$$= \frac{96{,}000}{319{,}510} = .30$$

REFERENCES AND RECOMMENDED READINGS

Adrian, J.L. and E.W. McCoy. 1972. *Experience and Location as Factors Influencing Income from Commercial Catfish Enterprises*. Bulletin 437, Agricultural Experiment Station/Auburn University, 28 pages.

Allen, P.G. and W.E. Johnston. 1976. "Research Direction and Economic Feasibility: An Example of Systems Analysis for Lobster Aquaculture." *Aquaculture*. 9:155-180.

Bardach, J.E. 1976. "Aquaculture Revisited." *Journal of the Fisheries Research Board of Canada*. 33,4(Part 2):880-7.

Bauer, L.L., P.A. Sandifer, T.I.J. Smith, and W.E. Jenkins. 1983. "Economics

Feasibility of Prawn *Macrobrachium* Production in South Carolina." *U.S.A. Aquaculture Engineering*. 2:181-201.

Blommestein, E., H. Deese, and J.P. McVey. 1977. "Socio-Economic Feasibility Studies of (*Macrobrachium rosenbergii*) Farming in Pala." *Journal of World Mariculture Society*. pp. 747-63.

Boyd, C.E., R.B. Rajendren, and J. Durda. 1986. "Economic Considerations of Fish Pond Aeration." *Journal of Aquacultural Tropics*. 1:1-5.

Chaston, I. 1984. *Business Management in Fisheries and Aquaculture*, England: Fishing News Books Ltd. 128 pages.

Chong, K., I.R. Smith, and M.S. Lizarondo. 1982. *Economies of the Philippine Milkfish Resource System*, Resource Systems Theory and Methodology Series, No. 4; The United Nations University. 66 pages.

Clonts, H.A. and S.B. Williams. 1983. "An Economic Assessment of Groundwater Use by Catfish Farmers in Alabama." *Midsouth Journal of Economics*. 7(3):457-64.

Crews, J. and J.W. Jensen. 1988. *Budget and Sensitivity Analysis for Alabama Catfish Production*, Agriculture Economic Series, Agriculture and Natural Resources, Alabama Cooperative Extension Service, Auburn University, Auburn, Alabama.

Engle, C.R. and U. Hatch. 1988. "Economic Assessment of Alternative Aquaculture Aeration Strategies." *Journal of the World Aquaculture Society*. 19(3):85-96.

Engle, C.R. 1987. "Analysis economic de la produccion comercial de la Tilapia, Colossoma, y Macrobrachium rosenbergii en mono y policultivo en Panama." *Revista Latino Americana de Acuicultura*. Lima-Peru. No. 33:7-44.

Escover, E.M., O.T. Salon, and I.R. Smith. 1987. "The Economics of Tilapia Fingerling Production and Marketing in the Philippines." *Aquaculture and Fisheries Management*. 18:1-13.

Flynn, J.B., N.R. Martin, and G.D. Hansen. 1983. *Effect of Production and Credit Management Factors on Catfish Investment and Profitability*, Bulletin 548. Alabama Agricultural Experiment Station, Auburn University, Auburn, Alabama. 27 pages.

Food and Agriculture Organization of the United Nations. 1987. *Aquaculture for Local Community Development Programme: Technical Consultation on Aquaculture in Rural Development*, FAO, Rome, Italy, 84 pages.

Gates, J.M. 1972. "Appraising the Feasibility of Fish Culture." *OECD Economic Aspects of Fish Production*, Paris, France, pp. 327-48.

Glude, B.J. 1966. "Criteria of Success and Failure in the Management of Shellfisheries." *Transactions of the American Fisheries Society*. 95:260-3.

Griffin, W.L., J.S. Hanson, R.W. Brick, and M.A. Johns. 1981. "Bioeconomic Modeling with Stochastic Elements in Shrimp Culture." *Journal of World Mariculture Society*. 12(1):94-103.

Hanson, J.S., W.L. Griffin, J.W. Richardson, and C.J. Nixon. 1985. "Economic Feasibility of Shrimp Farming in Texas: An Investment Analysis for

Semi-Intensive Pond Grow-out." *Journal of World Mariculture Society*. 16:129-50.

Hatch, U. and J. Atwood. 1988. "A Risk Programming Model for Farm-Raised Catfish." *Aquaculture*. 70:219-30.

Hatch, U. and C. Engle. 1987. "Economic Analysis of Aquaculture as a Component of Integrated Agro-Aquaculture Systems: Some Evidence from Panama." *Journal of Aquacultural Tropics*. 2:93-105.

Hatch, U., S. Sindelar, D. Rouse, and H. Perez. 1987. "Demonstrating the Use of Risk Programming for Aquacultural Farm Management: The Case of Penaeid Shrimp in Panama." *Journal of the World Aquaculture Society*. 18(4):260-9.

Hopkins, K., M. Hopkins, D. Lederig, and A. Al-Meeri. 1986. "Tilapia Culture in Kuwait: A Preliminary Economic Analysis of Production Systems." *Kuwait Bulletin of Marine Science*. 7:45-64.

Huang, H., W.L. Griffin, and D.V. Aldrich. 1984. "A Preliminary Economic Feasibility Analysis of a Proposed Commercial Penaeid Shrimp Culture." *Journal of World Mariculture Society*. 15:95-105.

Johns, M., W. Griffin, A. Lawrence, and D. Hutchins. 1981. "Budget Analysis of Shrimp Maturation Facility." *Journal of World Mariculture Society*. 12(1):104-9.

Jolly, C.M. and C.R. Engle. 1988. "Effects of Stocking, Harvesting, and Marketing Strategies on Profit Maximization in Catfish Farming." *The Southern Business and Economic Journal*. 12(1):52-62.

Klemetson, S.L. and G.L. Rogers. 1985. "Engineering and Economic Considerations for Aquaculture Development." *Aquaculture Engineering*. 4:1-9.

Leeds, R. 1986. "Financing Aquaculture Projects." *Aquacultural Engineering*. 5:109-13.

Liao, D.S., and T.I.J. Smith. 1983. "Economics Analysis of Small-Scale Prawn Farming in South Carolina." *Journal of World Mariculture Society*. 14:441-50.

Leopold, M. 1981. *Problems of Fish Culture Economics with Special Reference to Carp Culture in Eastern Europe*. FAO, Rome, Italy, 99 pages.

Lovshin, L.L., N.B. Schwartz, V.G. DeCastillo, C.R. Engle, and U.L. Hatch. 1986. *Cooperatively Managed Rural Panamian Fish Ponds: The Integrated Approach*. International Center for Aquaculture Research Development Series, No. 33. Alabama Agricultural Experiment Station, Auburn University, Auburn, Alabama, 47 pages.

McCoy, H.D. 1987. "Intensive Culture, the Past, the Present, and Future: Part III." *Aquaculture Magazine*. March/April, pp. 24-19.

Miyamura, M. and J. Katoh, 1986. "Fundamentals of Planning and Designing of Aquacultural Ponds." *Journal of Aquacultural Tropics*. 1:75-97.

Panayatou, T., S. Wattanuchariya, S. Isvilanonda, and R. Tokrisna. 1982. *The Economics of Catfish Farming in Central Thailand*, ICLARM Technical Reports 4. Manila, Philippines. 60 pages.

Pomeroy, R.S., D.B. Luke and T. Schwedler. 1987. "The Economics of Catfish

Production in South Carolina." *Aquaculture Magazine*. January/February. pp. 29-35.

Rabanal, H.R. 1987. "Managing the Development of Aquaculture Fisheries." *Journal of the World Aquaculture Society*. 18,2:117-25.

Rauch, H.E., L.W. Botsford, and R.A. Shleser. 1975. "Economic Optimization of an Aquaculture Facility." *IEEE Transactions on Automatic Control*. AC-20(3):310-9.

Rhodes, R.J. 1984. "Primer on Aquaculture Finances: Taxes, Time, and Investors, Part III." *Aquaculture Magazine*. March/April, pp. 22-5.

Rhodes, R.J. 1983. "Primer on Aquaculture Finances Planning for Success, Part I." *Aquaculture Magazine*. November/December, pp. 16-20.

Roberts, K.J. and L.L. Bauer. 1978. "Costs and Returns for *Macrobrachium* grow-out in South Carolina, U.S.A." *Aquaculture*. 15:383-90.

Samples, K.C. and P. Leung. 1985. "The Effect of Production Variability on Financial Risks of Freshwater Prawn Farming in Hawaii." *Canadian Journal of Fisheries Aquacultural Sciences*. 42:307-11.

Shang, Y.C. 1986. "Research on Aquaculture Economics: A Review." *Aquacultural Engineering*. 5:103-08.

Shang, Y.C. and W.J. Baldwin. 1980. "Economic Aspects of Pond Culture of Topminnows (*Family Poeciliidae*) in Hawaii as an Alternative Baitfish for Skipjack Tuna." *Processors World Mariculture Society*. 11:592-5.

Shang, Y.C. and T. Fujimura. 1977. "The Production Economics of Freshwater Prawn (*Macrobrachium rosenbergii*) Farming in Hawaii." *Aquaculture*. 11:99-110.

Shigekawa, K.J. and S.H. Logan. 1986. "Economic Analysis of Commercial Hatchery Production of Sturgeon." *Aquaculture*. 51:299-312.

Shlesser, R. 1986. "Aquaculture in Hawaii, A Social and Economic Dilemma." *Aquaculture Engineering*. 5:87-101.

Stamp, N.H.E. 1978. "Computer Technology and Farm Management Economics in Shrimp Farming." *Journal of World Mariculture Society*. 9:383-91.

Stokes, R.L. 1982. "The Economics of Salmon Ranching." *Land Economics*. 58(4):464-77.

Varley, R.L. 1977. "Economics of Fish Farming in the United Kingdom." *Fish Farming International*. pp. 17-9.

Walker, N.P. and J.M. Gates. 1981. "Financial Feasibility of High Density Oyster Culture in Saltmarsh Ponds with Artificially Prolonged Tidal Flows." *Aquaculture*. 22:11-20.

Chapter 8

The Time Value of Money

Fish production, like many economic decisions, involves benefits and costs that are expected to occur at future time periods. The construction of ponds, raceways, and fish tanks, for example, requires immediate cash outlay, which, with the production and sale of fish, will result in future cash inflows or returns. In order to determine whether the future cash inflows justify present initial investment, we must compare money spent today with money received in the future.

PRESENT VALUE VERSUS FUTURE VALUE

Can we successfully compare money spent today with money we will receive a year from now, 2 years from now, or 10 years from now? Is $1.00 received today equivalent to $1.00 received a year from today? The value of the dollar a year from today depends on the alternative use available for the dollar between today and a year from today. Suppose instead of investing the dollar in finfish production, which yields an interest rate of 6.0 percent, we invest it in shellfish production which yields 10 percent per annum. One year from now, if we invested in finfish, we should receive $1.06 whereas if we invested in shellfish, we would receive $1.10. The difference is the result of the interest rate, or the rate of return on the investment. The possibility of investing in a project at a positive interest rate changes the value of the investment. If we chose to lend someone the dollar instead of investing it in finfish production, and a year from now only the dollar note is returned, then the dollar is worth only $0.94 today if it is worth $1.00 next year. This is because of *time value of money*. The process of finding present value is called *discounting*.

The time value of money influences many production decisions. Everyone prefers money today over money in the future. Therefore, in order to invest a dollar in fish production today, one must be guaranteed a return in the future that is equal to or greater than the dollar invested today. The preference for the dollar now instead of a dollar in the future arises from three basic reasons: uncertainty, alternative uses, and inflation. Uncertainty influences preference because one is never sure what will take place tomorrow. The alternative uses will determine whether one invests in one project or another, and inflation affects the purchasing power of the dollar.

The interest rate (i) is considered an exchange price between present and future dollars. Thus, $1.00 today exchanges for $1 + i$ dollars one period in the future. Or alternatively, a $1.00 payment-made one period in the future exchanges for $1/1 + i$ dollars now. Interest rates equate present and future claims for financial assets of different maturities. These rates respond to changes in supply and demand for alternative financial assets, including money, just as other commodity prices respond to changes in their supply and demand. Interest rates are always positive because of the positive time preference for money. The positive time preference means the sooner money is available, the greater its value.

FUTURE VALUE OF A PRESENT SUM – COMPOUNDING

In the example given earlier, one time period was examined, that is, 1 year. If a dollar were invested today at an interest rate of 6.0 percent per annum, what will be the value 2 years hence? Would it be $1.20 or $1.12? The value $1.12 is based on the simple interest rate; that is, only the original amount of money ($1.00) earns interest over the 2 years. However, in real business, interest is earned on the original amount plus the interest income. At the end of the transaction period, the total principal available is called the *compound amount*, and the difference between the original principal and the compound amount is called *compound interest*. The rate of interest is called the *compound interest rate*.

Suppose you have \$1,000 to invest in a bank paying interest at 6.0 percent compounded annually ($i = 6\%$). After 1 year you will have:

$$\$1{,}000 + (1{,}000)(i), \text{ or}$$
$$1{,}000\,(1 + i) = 1{,}000(1.06) = 1{,}060$$

After 2 years you will have:

$$1{,}000(1 + i)(1 + i) = 1{,}000(1.06)(1.06) = 1{,}123.60, \text{ or}$$
$$1{,}000(1 + i)^2 = 1{,}123.60$$

The \$1,123.60 is the compound amount or future value of the \$1,000 principal, or present value, invested at a 6.0 percent compound interest rate for 2 years. The compound interest earned is \$123.60.

A general formula for obtaining the future of a present sum may be written as:

$$V_N = V_0\,(1 + i)^N$$

where:

V_N = future value
V_0 = present value
i = interest rate
N = number of conversion periods.

Consider an example. Suppose we have \$1,500 and we would like to invest it today in fish ponds which will yield a return of 10 percent per annum. What will be the value 5 years from now? The future value may be calculated using the formula:

$$V_N = V_0\,(1 + i)^N$$

where:

$$V_0 = \$1{,}500$$
$$i = .10$$
$$N = 5$$

therefore:

$$V_N = 1{,}500\,(1 + .10)^5$$
$$= 1{,}500\,(1.610)$$
$$= \$2{,}415$$

Consider another example. Suppose we have just sold our fish at the end of November and our profit is $20,000. The bank just sent a notice saying that it offers an interest rate of 9.0 percent on money market certificates, if $20,000 or more is invested for a period of no less than 5 years. What will be the value of $20,000 5 years from now? For this calculation:

$$V_0 = \$20{,}000$$
$$i = .09$$
$$N = 5$$

therefore:

$$V_N = 20{,}000\,(1.09)^5$$
$$= 20{,}000\,(1.5386)$$
$$= \$30{,}772$$

If the money is invested in money market certificates, the earnings will be $30,772 over 5 years. This would be acceptable only if it was a greater amount than could be earned in some alternative use, as in this case, growing fish.

Future Value of a Series of Payments

Sometimes farmers invest in an enterprise which will produce a series of payments. Assume that a fish farmer wants to invest money in a catfish production activity which will generate returns over a number of years (N). The fish farmer wants to know the value of the payments or returns after a number of years, say upon retirement age. We are finding the future value of a series of payments which is easy to calculate using the formula:

$$V_N = P_0(1 + i)^N + P_1(1 + 1)^{N-1} + P_2(1 + i)^{N-2} \ldots + P_n$$

$$= \sum_{N=0}^{N} P_N(1 + i)^{N-n}$$

where:

V_N = the future value of the series of payments
P_N = the payment for each conversion period (n) and (n = 0, 1, 2,...N).

Next, consider the income from an aerator which will yield income flows of $300, $400, $500, $600, and $700 during the 1st to 5th year of functioning. If we assume that the interest rate is 9.0 percent, what is the future value of this series of payments?

$$V_N = \$300(1.09)^4 + 400(1.09)^3 + 500(1.09)^2 + 600(1.09) + 700$$

$$V_N = 423.47 + 518.01 + 594.05 + 654.00 + 700.00$$

$$= \$2,889.53$$

Note that the summed value of the income generated over the 5-year period is $2,500. The additional accrued amount is the result of compounding, since it is believed that the income received is invested at 9.0 percent per annum.

Future Value of a Uniform Series

If the returns were uniform, the following formula could be used:

$$V_N = A\left[\frac{(1 + i)^N - 1}{i}\right]$$

where:

A = the annuity payment in each period
N = the number of periods
i = the interest rate.

If the annuity for the aerator was $500 then the future value would be:

$$V_n = 500 \left[\frac{(1.09)^5 - 1}{.09} \right] = 500(5.9847) = 2,992.36$$

PRESENT VALUE OF A FUTURE PAYMENT–DISCOUNTING

The present value is found by solving the previous equation for V_0 as opposed to V_N. The result is a general formula for determining the present value of a future sum:

$$V_0 = \frac{V_N}{(1 + i)^N}$$

or

$$V_0 = V_N(1 + i)^{-N}$$

The future value must be *discounted* to reflect the earnings lost by not being able to immediately invest the future sum in the alternative investment, yielding interest rate (i). *Discounting* does not mean that future earnings are actually "lost," but that they are not worth the same in the present as in the future because of risk and uncertainty regarding future events. From our previous example (the sale of fish), if we want to find out the present value of $30,772 available 5 years from now at an interest rate of 9 percent, we may use the formula:

$$V_0 = \frac{V_N}{(1 + i)^N}$$

where:

$$V_N = \$30{,}772$$
$$i = .09$$
$$N = 5$$

therefore:

$$V_0 = \frac{\$30{,}772}{(1.09)^5}$$

$$= \frac{30{,}772}{1.5386} = \$20{,}000$$

Consider another example. Suppose that we are offered a job as a manager of a fish farm in Indonesia after the completion of our studies. The salary is low, but the offer includes a gratuity payment of $15,000(U.S.) at the end of a 5-year contract. Assuming the discount rate is 12 percent, what is the present value of this lump sum gratuity payment?

$$V_0 = \frac{V_N}{(1 + i)^N}$$

where:

$$V_N = \$15{,}000$$
$$i = .12$$
$$N = 5$$

therefore:

$$V_0 = \frac{\$15{,}000}{(1+.12)^5} = \frac{\$15{,}000}{1.7623} = \$8{,}511.60$$

Present Value of a Series of Payments

Investments in aquacultural enterprises frequently generate a series of payments over the life of the investment, although the initial outlay may be a one-time cash expenditure. Let's say we would like to invest in an aerator which would result in added production of fish for 5 years at a discount rate (i). We would like to find out the present value of the future series of payments. The procedure for determining the present value of a series of payments is simply an extension of the method for determining the present value of a single future sum. Each future payment is discounted in its respective year. The formula is:

$$V_0 = \frac{P_1}{(1+i)} + \frac{P_2}{(1+i)^2} + \frac{P_3}{(1+i)^3} \cdots + \frac{P_N}{(1+i)^N}$$

or

$$V_0 = \sum_{n=0}^{N} \frac{P_n}{(1+i)^n}$$

where:

V_0 = the present value of the payment series
P_N = the payment for each conversion period (n) and (N = 0, 1, 2,. . .N)
i = the interest rate
Σ = the summation for n = 0 to N.

For the aerator investment mentioned earlier, if the return for 5 years is $300, $400, $500, $600 and $700 for the 1st, 2nd, 3rd, 4th and 5th year respectively at a discounted rate of 9 percent, the present value of the returns is:

$$V_0 = \frac{\$300}{1.09} + \frac{\$400}{(1.09)^2} + \frac{\$500}{(1.09)^3} + \frac{\$600}{(1.09)^4} + \frac{\$700}{(1.09)^5}$$

or we can use the formula:

$$V_0 = P_1(1 + i) + P_2(1 + i)^{-2} \ldots + P_n (1 + i)^{-N}$$
$$V_0 = 300(.9174) + 400(.8417) + 500(.7722)$$
$$+ 600(.7084) + 700(.6499)$$
$$= 275.22 + 336.68 + 386.10 + 425.04 + 454.93$$
$$= \$1,877.97$$

Present Value of a Uniform Series

If the future payments are equal over the period, we can use the present value formula for a uniform series, which is stated as:

$$V_0 = A\left[\frac{1-(1 + i)^{-N}}{i}\right]$$

where:

A = annuity payment in each period
i = interest rate
N = number of periods.

This formula is sometimes written as:

$$V_0 = A\left[\frac{1-(1 + i)^{-N}}{i}\right]$$

or as:

$$V_0 = A[USPV_i,N]$$

where:

$USPV_i,N$ represents uniform series present value for N periods at interest rate i.

For example, assume that the returns on the aerator were $500 for 5 years at 9 percent per annum. The present value of the series would be:

$$V_0 = \left[500 \frac{1-(1.09)^{-5}}{.09}\right] = 500\ [3.9927]$$

$$= 1{,}995$$

Present Value of an Infinite Uniform Series

In aquacultural production, certain assumptions are often made when constructing a pond or some of the structures related to fish production. First, it is frequently assumed that the pond will last indefinitely. The same assumptions are made for drains and other holding tanks. These assumptions are made because these structures are considered to be nondepreciable assets. If assets are nondepreciable, we would expect them to produce returns indefinitely. An examination of the formula for present value of a uniform series will show that as N increases for a given annuity, the present value increases at a decreasing rate. Each period adds a discounted payment to the income stream. This continues until the last annuity approaches zero. Therefore, the equation may be represented as:

$$V_0 = (1 + i)^{-N}$$

or

$$V_0 = \frac{A}{i}$$

This formula is used for the valuation of land and other assets which will bring an indefinite flow of returns on investment.

Compounding Period

In the previous discussion, it was assumed that the rate of return (i) was compounded annually. In many business dealings, while the rate is expressed as an annual rate, compounding occurs more frequently, sometimes on a monthly, quarterly, or semi-annual basis. The more times per year the return is compounded, the larger will be the total return and the smaller the present value. Under these circumstances, the interest rate must be changed from an annual rate to one per conversion period. To do this, the following formula is used:

$$V_N = V_0 \left(1 + \frac{i}{m}\right)^{mN}$$

where:

m = number of conversion periods per year
N = number of years.

For example, suppose the bank is offering a 10 percent annual rate of interest on our $20,000 with interest compounded quarterly (every 3 months). The number of conversion periods is 4 and the interest rate per period is 2.5 percent. By investing $20,000 for 10 years we will have:

$$\begin{aligned} V_n &= 20{,}000 \left(1 + \frac{.04}{4}\right)^{40} \\ &= 20{,}000\ (2.435) \\ &= 48{,}700 \end{aligned}$$

REFERENCES AND RECOMMENDED READINGS

Barry, P.J., A.J. Hopkin, and C.B. Baker. 1979. *Financial Management in Agriculture*. Fourth Edition. Illinois: The Interstate Printers and Publishers, Inc. 529 pages.

McGuigan, J.R. and R.C. Moyer. 1989. *Managerial Economics*. St. Paul: West Publishing Company. 653 pages.

Warren, L.F, M.D. Boehlje, A.G. Nelson, and W.G. Murray. 1988. *Agricultural Finance*. Eighth Edition, Ames: Iowa State University Press. 468 pages.

Weston, F.J. and E.F. Brigham, 1977. *Managerial Finance*. Sixth Edition. 1030 pages. Illinois: The Dryden Press.

Chapter 9

Capital Budgeting

The aquaculturist is often confronted with the problem of evaluating projects that will last several years and that have varying project costs and benefits over the life of the project. It is not always easy to choose one project over another when the projects vary in size. The method most commonly used for addressing this problem is capital budgeting. Capital budgeting is a process of evaluating the profitability of projects that involve large sums of money whose returns are expected to extend beyond a year.

There are several examples of capital outlays in aquaculture which require large sums of money and whose costs and returns extend over a long period. Examples are pond construction, farm buildings, tractors, water reservoirs, roads, processing equipment, and other capital investments associated with plant expansion. Advertising campaigns or research and development projects may also be classified as capital budgeting expenditures because they are also likely to have an impact beyond a year. Capital budgeting involves the entire planning of the project expenditures whose returns occur for a period of more than a year. The process of budgeting should follow an orderly sequence of steps that produce information relevant to an investment choice. These steps are generally classified as:

1. identifying alternative projects;
2. specifying relevant criteria for choosing among alternatives;
3. data collection on selected alternatives;
4. analysis of relevant data; and
5. data interpretation and final project selection.

The fish farmer or decision maker will search for the fish enterprise that will produce the most net benefits for him and the farm family. A project manager will search for the project that will satisfy the project goals, and that includes the wise and best use of resources to satisfy human wants. The most common criteria employed in determining the financial desirability of investment projects are:

1. payback period;
2. internal rate of return (IRR);
3. net present value (NPV); and
4. profitability index.

Payback Period

The payback period is a simple method which estimates the length of time required for an investment to pay itself out; that is, the number of years required for a firm to cover its original investment from net cash inflows. Consider two projects, A and B, shown in Table 9-1.

The first project (A) is a $2,000 investment for the purchase of one aerator, and the second (B) is to invest in a feed shed of equal cost. The payback period for the aerator is 3.5 years and that for the feed shed is approximately 4.3 years. If the decision maker wants to cover the cost of investment in the shortest period of time, project A, investment in an aerator, would be chosen over project B, the feed shed.

TABLE 9-1. Net Cash Inflows for Projects A and B

Year	A	B
0	-$2,000	-$2,000
1	700	300
2	600	400
3	500	500
4	400	600
5	300	700
6	200	800
7	--	900
8	--	1000

Although the payback period is easy to calculate, it can lead to erroneous decisions. As can be seen from our example, it ignores income beyond the payback period, and therefore is biased towards projects with shorter maturity periods.

The payback period is sometimes used by investors who are short of cash and need to reinvest all cash flows that occur in early stages of the project. Investors who are risk averse often use this technique in evaluating projects. Such investors need to receive cash at the early stages of the project since the future is uncertain. Thus, the payback method is a somewhat better reflection of liquidity than profitability.

NET PRESENT VALUE

The net present value (NPV) is a discounted cash flow technique (DCF). It is the present value discounted at the firm's required rate of return on the stream of net cash flows from the project *minus* the project's net investment. The NPV method uses the discounting formulas of a nonuniform or uniform series of payments to value the projected cash flows for each investment alternative at one point in time. To obtain the NPV, the following formula is used:

$$NPV = -INV + \frac{P_1}{(1+i)} + \frac{P_2}{(1+i)^2} + \frac{P_3}{(1+i)^3} \ldots + \frac{P_n}{(1+i)^N}$$

where:

$P_1 \ldots P_n$ are net cash flows
i = the interest rate or marginal cost of capital
N = the project expected life
INV = the initial investment.

The model indicates that the net cash flows of the project are discounted and then added to yield the NPV. Any salvage or terminal value is included as a cash flow in the last year of the project. The initial investment is negative since it represents a cash outflow.

In using NPV as a choice criterion between two projects, consider the investment in A, the purchase of a small tractor, or B, the investment in a small truck. Let us assume that driving conditions are severe and the vehicles are estimated to last for a period of 5 years. The inflows and present values of the inflows are shown in Table 9-2. The cost of capital is 10 percent.

Project A has an NPV of $2,100 and project B, $640. The difference in NPV is due to the time in which the funds are generated. Project A generates more returns during its early life, whereas larger returns are generated late in the life of project B. Returns further into the future are more uncertain. Thus, they are discounted more heavily.

If the projects generated equal returns (annuity) throughout the project life, then the formula for determining the NPV would be:

$$NPV = \sum_{t=1}^{n} \frac{P}{(1 + i)^t} - INV$$

If both projects generated an average $3,000 per annum, then the NPV would be obtained by using the net present value of a uniform series. Use of the tables in Appendix III is necessary to find the present value of $1.00 at, say, 10 percent for 5 years.

$$NPV = \sum_{1}^{5} \frac{3,000}{(1 + .10)^5} - 10,000$$

$$\begin{aligned} NPV &= 3,000\,(3.79) - 10,000 \\ &= 11,370 - 10,000 \\ &= 1,370 \end{aligned}$$

An investment project would be accepted if the NPV > 0, and rejected if NPV < 0. This is because the money being invested is greater than the present value of the net cash flow. If NPV = 0, the decision maker would be indifferent. If a project were cost-reducing, for example building concrete dikes, then projected cash flows

TABLE 9-2. Calculation of Net Present Value of Projects A and B with $10,000 Investment Cost

		Project A			Project B	
Year	Net cash flows	Discount rate	Present value	Net cash flow	Discount rate	Present value
1	5,000	.91	4,550	1,000	.91	910
2	4,000	.83	3,320	2,000	.83	1,660
3	3,000	.75	2,250	3,000	.75	2,250
4	2,000	.68	1,360	4,000	.68	2,720
5	1,000	.62	620	5,000	.62	3,100
			12,100			10,640

would reflect cash outlays. When cost-reducing investments are compared, the choice criterion is based on the minimum net present value of cash outlays.

INTERNAL RATE OF RETURN

The internal rate of return (IRR) is the interest rate that equates the present value of the expected future cash flows, or receipts, to the initial investment (INV) or cost outlay. To find the IRR for a project, the INV is *subtracted from or equated to* the net cash flows at a given interest rate. The formula is then:

$$INV = \frac{P_1}{(1+i)} + \frac{P_2}{(1+i)^2} + \ldots + \frac{P_N}{(1+i)^N} + \frac{V_N}{(1+i)^N}$$

What is sought is an interest rate that will *equate* the sum of the net cash flows to the initial investment. The interest rate that satisfies the equation is called the *internal rate of return* (IRR). The formula can thus be written:

$$0 = -INV + \frac{P_1}{(1+i)} + \frac{P_2}{(1+i)^2} + \ldots + \frac{P_n}{(1+i)^N} + \frac{V_N}{(1+i)^N}$$

where:

V_N is the end or salvage value of the investment.

For example, find the IRR of two projects: A, the investment in a small tractor which requires an initial outlay of $10,000; and B, the investment in a half-ton truck which is used for transporting feed, also costing $10,000. The investment in projects A and B yields the returns shown in Table 9-3.

There is no *one* way of finding the IRR. One is forced to use a systematic procedure of trial and error to find the discount rate that will equate the NPV to the initial investment. The most difficult aspect of trial and error is making the initial estimate. The following series of steps is given which will facilitate the process:

TABLE 9-3. Returns from Two Projects, A and B

Year	Project A	Project B
1	$5,000	1,000
2	4,000	2,000
3	3,000	3,000
4	2,000	4,000
5	1,000	5,000
	15,000	15,000

1. Divide the cash investment by the average of annual returns.

$$\text{Average returns: } \frac{\$15,000}{5} = 3,000$$

$$\text{Project A: } \frac{\$10,000}{3,000} = 3.33$$

$$\text{Project B: } \frac{10,000}{3,000} = 3.33$$

2. Use the present value of annuity table in Appendix III, Table IV because the varying annual returns are represented by a single average return. Look under years column and locate the number that the annual net cash returns are expected to last, 5 in this case.
3. Along the 5-year row, locate a factor equal or close to the quotient, 3.33.
4. Go to the top of the column. The interest rate or discount rate on top of the column is merely a first approximation of the true rate of return. The factor that closely approximates 3.33 is 3.352 under 15%.
5. If the annual net cash return during the earlier year is greater than that of the later year, move *at least* two interest rate percentages ***higher*** than the located interest rate. Skip two columns in this case to find 18%, which is the first trial rate.

6. If the annual net cash return during the earlier year is less than that of the later year, move *at least* two interest rate percentages *lower* than the located interest rate. Skip two columns in this case to find 12%, which is the first trial rate on the low side.
7. Use the present value table because the series of annual net cash returns is uneven. Multiply each annual net cash return by the yearly factor listed under the interest rate being tested. This is shown in Table 9-4.

If the total discounted net cash returns do not approximate the amount of investment, another trial is made until an interest rate is found that will equate or approximate the discounted net cash returns to the amount of investment. Often, an exact interest rate cannot be found in the tables. One must accept the rate closest to the true rate.

The calculations in Table 9-4 indicate that the approximated internal rate of return for Project B, which is 12 percent, is the true rate because the difference between the discounted cash return and the amount of investment, $2, is insignificant.

The first approximated rate of 18 percent for Project A is not the

TABLE 9-4. Calculation of Internal Rate of Return for Projects A and B

	Project A			Project B		
Year	Net cash return	Factor (18% - 5 yrs)	Discounted return	Net cash return	Factor (12% - 5 yrs)	Discounted return
1	$5,000	0.847	$ 4,235	$ 1,000	0.893	$ 893
2	4,000	0.718	2,872	2,000	0.797	1,594
3	3,000	0.609	1,827	3,000	0.712	2,136
4	2,000	0.516	1,032	4,000	0.636	2,544
5	1,000	0.437	437	5,000	0.567	2,835
Total discounted net cash returns			$10,403			$10,002

			Project A	Project B
Total discounted returns		$10,403	$10,002	
Amount of investment	10,000	10,000		
Excess			$ 403	$ 2

true rate because the discounted cash returns are greater than the amount of investment by $403. The true rate should be higher than 18 percent. Using a second approximated rate of 20 percent, the discounted net cash returns for Project A are calculated in Table 9-5.

The approximated internal rate of return of 20 percent for Project A is considered as the true rate because the excess of $44 is negligible.

The IRR may be used for ranking projects. Between the two projects in this example, project A is more desirable. The ranking is based on the relative size of the IRR, with the largest IRR receiving the highest rank. Acceptability of each project depends upon comparing the IRR with the investor's *required rate of return* (RRR), sometimes called *minimum acceptable rate of return* (MARR). If IRR is greater than the RRR (MARR), accept the project; if IRR is less, reject the project; if IRR equals the required rate, be indifferent. For our two projects A and B, if the required rate were 10 percent, then both projects would be acceptable, though project A would be more desirable.

Both the NPV and the IRR methods result in identical decisions to either accept or reject an independent project. This is true because the NPV is greater than (less than) 0 if, and only if, the IRR is greater (less than) the RRR. In mutually exclusive projects, the two

TABLE 9-5. Calculation of Internal Rate of Return for Project A

Year	Net Cash returns	Factor (20% - 5 yrs)	Discounted returns
1	$ 5,000	0.833	$ 4,165
2	4,000	0.694	2,776
3	3,000	0.579	1,737
4	2,000	0.482	964
5	1,000	0.402	402
Total discounted net cash returns			$10,044
Amount of investment			10,000
Excess			$ 44

methods may yield contradictory results. The IRR method implicitly assumes that returns from an investment are reinvested to earn the same rate as the IRR of the investment. The NPV method, on the other hand, assumes that these funds may be reinvested at the firm's interest rate.

Profitability Index

The profitability index (PI) is sometimes referred to as the benefit-cost ratio. It is the ratio of the present value of future net cash flows over the life of the project to the net investment. The index is represented as:

$$\frac{\sum_{t=1}^{N} \frac{P}{(1+i)^t}}{INV}$$

This method usually produces the same result as the NPV and IRR in project evaluation, but it is very important in separating projects of varying sizes. If a project has a PI greater than or equal to 1, it should be accepted, and should be rejected if the PI is less than 1. In the example of our two projects, A, the investment in a small tractor, and B, the investment in a small truck, the PIs are calculated as:

$$\text{PI for project A} = \frac{12{,}100}{10{,}000} = 1.2$$

$$\text{PI for project B} = \frac{10{,}640}{10{,}000} = 1.06$$

Both the PIs are greater than 1, therefore both projects are acceptable although the PI for project A is larger.

When there is substantial difference in sizes of investments, there may be a difference in the results of NPV, IRR, and PI. When there is a conflict, other factors must be used for project evaluation.

Projects with Investments at Different Periods

The investment projects used for illustration were simple one-time investments, but most long-term projects require capital renewal at different times of the project life. Variable time investments may be illustrated with a single project which involves the investment in a 1-acre tilapia farm. The assumption is made that the project life is 5 years. Table 9-6 shows both investment and operating costs as well as cash flows for the 5 years of project life. The net returns or net cash flows are obtained by subtracting total costs from gross revenue.

To determine the feasibility of the project, it is necessary to calculate the Net Present Value and the Internal Rate of Return. To do this, Table 9-6 is reorganized and shown in Table 9-7.

The internal rate of return may be computed by the trial and error method or by the use of a computer package. We will use the trial and error approach. We know that the IRR will be greater than 12 percent since the NPV > 0 at 12 percent. Let's try 25 percent. Go to Appendix Table II and look up the discount factors at 25 percent for 5 years and then calculate the IRR:

$$\begin{aligned} \text{IRR} &= (-6{,}400 \times .800) + (1{,}200 \times .64) + (3{,}000 \times .512) \\ &+ (3{,}200 \times .400) + (4{,}200 \times .328) \\ &= -5{,}120.00 + 768.00 + 1{,}536.00 + 1{,}312.00 + 1{,}377.60 \\ &= -126 \end{aligned}$$

Since −126 is large, try 24 percent:

$$\begin{aligned} \text{IRR} &= (-6{,}400 \times .806) + (1{,}200 \times .650) + (3{,}000 \times .524) \\ &+ (3{,}200 \times .423) + (4{,}200 \times .34) \\ &= -5{,}158.40 + 780.00 + 1{,}572.00 + 1{,}353.60 + 1{,}432.20 \\ &= -20.60 \end{aligned}$$

Since −20.60 is close enough to 0, the IRR is approximately 24 percent.

The profitability index (PI) may now be calculated using the data in Table 9-8 (and: total investment = $7,900 + 300 (.64) = $8,092 from Table 9-6):

TABLE 9-6. Illustration of the Investment for a 1-Acre Tilapia Pond for 5 Years

Initial capital costs (in U.S. $)	Year 1	2	3	4	5
Pond construction	5,000	-----	-----	-----	-----
Well	1,000	-----	-----	-----	-----
Office and Storage	1,000	-----	-----	-----	-----
Pump	300	-----	-----	300	-----
Nets	300	-----	-----	-----	-----
Miscellaneous	300	-----	-----	-----	-----
Sub-Total	7,900	-----	-----	300	-----
Operating Costs (excluding interest and depreciation)	1,500	3,800	4,000	4,500	4,800
Gross revenue/ year	3,000	5,000	7,000	8,000	9,000
Net Return per year	-6,400	1,200	3,000	3,200	4,200

Total investment equals initial costs plus additional cost in later years, or in this case $7,900 + 300 = $8,200.

$$PI = \frac{9{,}059}{8{,}092} = 1.12$$

The project is acceptable at an interest rate of 12 percent since the PI > 1.

The calculation of NPV, IRR, and PI for large or long-term investment projects may be somewhat laborious. A computer may be used to make calculations for any reasonably large project. Consider the financial analysis for 20 acres of catfish production in Alabama in 1990. The cost and return estimates are derived from recommended practices as stipulated in a budget analysis by Crews and Jensen (1989). Production is based on moderate hill pond management practices, which means that catfish are stocked in the

TABLE 9-7. Calculation of Net Present Value and Internal Rate of Return for a 1-Acre Tilapia Farm, U.S.$

Year	Revenue	Calculation of net present value Cost	Return	Rate at 12% discount	Discounted values Revenue	Cost	Net return
1	2	3	4	5	(2x5)	(3x5)	(4x5)
	(US $)	(US $)	(US $)	(US $)	(US $)	(US $)	(US $)
1	3,000	9,400	-6,400	0.89	2,670	8,366	-5,696
2	5,000	3,800	1,200	0.80	4,000	3,040	960
3	7,000	4,000	3,000	0.71	4,970	2,840	2,130
4	8,000	4,800	3,200	0.64	5,120	3,072	2,048
5	9,000	4,800	4,200	0.57	5,130	2,736	2,394
	32,000	26,800	5,200		21,890	20,054	1,836

Net present value = US $1,836

TABLE 9-8. Calculation of Profitability Index for a 1-Acre Tilapia Pond

Year	Net return	Discount rate (12%)	Net present value
1	1,500	.89	1,335
2	1,200	.80	960
3	3,000	.71	2,130
4	3,500	.64	2,240
5	4,200	.57	2,394
			9,059

spring at 3,500 fish per acre and ponds are drained in the fall for harvesting. A 6.0 percent death loss is anticipated. The production period is 200 days and the feed conversion ratio is 1.7 pounds of feed to 1.0 pound of gain. Fish are grown to 1.0 pound and sold at $.70 per pound to the processing plant. The life of the project is 20 years. Initial capital investment is made during the first year. Fish sales begin in the second year. There is some capital replacement during the fifth, tenth, and fifteenth years of the project.

The receipts, cash outflows, and net cash inflows are shown in Table 9-9. Pond construction was assumed to cost $1,500 an acre and all initial expense would come from equity capital. Also, it was assumed that input and output prices as well as the production level would be constant throughout the life of the project. The NPVs were obtained at 12.0 and 8.0 percent. The NPV was $16,149 at 8.0 percent and $ – 12,324 at 12.0 percent. The IRR was 10.0 percent. This means that capital should be borrowed at no more than 10.0 percent per annum to engage in this project, given the assumptions made, since the IRR is 10.0 percent.

Several assumptions were made in determining NPV and IRR (Table 9-10). For example, it was assumed that a 20-acre pond would be constructed, but the pond could be sold at the original price of construction at the end of the project. At 8.0 percent discount, the NPV was $22,586 and at 12.0 percent the NPV was $ – 9,215. The IRR was 10.6 percent. This means that the value of the pond 20 years in the future does not affect the IRR.

When pond construction cost was excluded from the calculation, the NPV at 8.0 percent was $43,927 and at 12.0 percent it was $14,461. The IRR was then 14.98 percent. Another variation shown in Table 9-10 is that instead of a single capital outlay of $30,000 for pond construction, payments of $1,500 would be deducted annually. The results in Table 9-10 show the NPV at 8.0 percent of $30,588 and $4,596 at 12.0 percent. The IRR was 12.9 percent. The information from Table 9-10 shows that the time and amount of capital outlay affected the NPV and IRR. The data also indicates that money received in the future, for example the receipts from sale of salvaged investments, is worth less than its present value.

TABLE 9-9. Projection of the Net Cash Flows for a 20-Acre Pond, 20 Years

Item	1	2	3	4	5	6	7	8	9
1)Operating Receipts		46060.00	46060.00	46060.00	46060.00	46060.00	46060.00	46060.00	46060.00
2)Terminal Value									
3)Total Cash Inflow (Line 1+2)		46060.00	46060.00	46060.00	46060.00	46060.00	46060.00	46060.00	46060.00
4)Initial Outlay	-117851.40								
5)Operating Expense		26579.81	26579.81	26579.81	26579.81	26579.81	26579.81	26579.81	26579.81
6)Depreciation		2497.00	2497.00	2497.00	2497.00	2491.00	2491.00	2491.00	2491.00
7)Recurrent Cost					975.00				
8)Taxable Income (Line 1-5-6-7)		16983.19	16983.19	16983.19	16008.19	16983.19	16983.19	16983.19	16983.19
9)Income Taxes (Line 8 x .28)		4755.29	4755.29	4755.29	4482.29	4755.29	4755.29	4755.29	4755.29
10)Total Cash Outflow (Line 4+5+7+9)	-117851.40	31335.10	31335.10	31335.10	32037.10	31335.10	31335.10	31335.10	31335.10
11)Net Cash Flow (Line 3-10)	-117851.40	14724.90	14724.90	14724.90	14022.90	14724.90	14724.90	14724.90	14724.90

TABLE 9-9 (continued)

	10	11	12	13	14	15	16	17	18	19	20
1	46060.00	46060.00	46060.00	46060.00	46060.00	46060.00	46060.00	46060.00	46060.00	46060.00	46060.00
2											
3	46060.00	46060.00	46060.00	46060.00	46060.00	46060.00	46060.00	46060.00	46060.00	46060.00	46060.00
4											
5	26579.81	26579.81	26579.81	26579.81	26579.81	26579.81	26579.81	26579.81	26579.81	26579.81	26579.81
6	2491.00	2491.00	2491.00	2491.00	2491.00	2491.00	2491.00	2491.00	2491.00	2491.00	2491.00
7	13806.75					2505.75					
8	3182.44	16983.19	16983.19	16983.19	16983.19	14483.44	16983.19	16983.19	16983.19	16983.19	16983.19
9	891.08	4755.29	4755.29	4755.29	4755.29	4055.36	4755.29	4755.29	4755.29	4755.29	4755.29
10	41277.64	31335.10	31335.10	31335.10	31335.10	33140.92	31335.10	31335.10	31335.10	31335.10	31335.10
11	4782.36	14724.90	14724.90	14724.90	14724.90	12919.08	14724.90	14724.90	14724.90	14724.90	14724.90

TABLE 9-10. Net Cash Inflows, NPV, and IRR for 4 Scenario for a 20-Acre Catfish Project in Alabama, 1989[a]

	1	2	3	4
	With Pond Construction	With Pond Construction Pond sold at original cost at end of project	Without Pond Construction	With Pond Construction with production payments of pond made annually
Year	Cash Flow	Cash Flow	Cash Flow	Cash Flow
1.00	-117851.40	-117851.40	- 87851.40	- 87851.40
2.00	14724.90	14724.90	14724.90	13224.90
3.00	14724.90	14724.90	14724.90	13224.90
4.00	14724.90	14724.90	14724.90	13224.90
5.00	14021.22	14021.22	14021.22	12521.22
6.00	14724.90	14724.90	14724.90	13224.90
7.00	14724.90	14724.90	14724.90	13224.90
8.00	14724.90	14724.90	14724.90	13224.90
9.00	14724.90	14724.90	14724.90	13224.90
10.00	4782.36	4782.36	4782.36	3282.36
11.00	14724.90	14724.90	14724.90	13224.90
12.00	14724.90	14724.90	14724.90	13224.90
13.00	14724.90	14724.90	14724.90	13224.90

[a] Estimations were derived from the catfish budget produced by Crews and Jensen, 1989.

TABLE 9-10 (continued)

	1	2	3	4
	With Pond Construction	With Pond Construction Pond sold at original cost at end of project	Without Pond Construction	With Pond Construction with production payments of pond made annually
Year	Cash Flow	Cash Flow	Cash Flow	Cash Flow
14.00	14724.90	14724.90	14724.90	13224.90
15.00	12919.08	12919.08	12919.08	11419.08
16.00	14724.90	14724.90	14724.90	13224.90
17.00	14724.90	14724.90	14724.90	13224.90
18.00	14724.90	14724.90	14724.90	13224.90
19.00	14724.90	14724.90	14724.90	13224.90
20.00	14724.90	14724.90	14724.90	13224.90
Total	149471.34	179471.34	179471.34	150971.34
NPV 12%	-12324.568	- 9214.5658	14461.145	4596.265
NFV 08%	16149.4153	22585.8615	43927.193	30588.860
IRR	0.10017690	0.103	0.1498	0.1297

[a]Estimations were derived from the catfish budget produced by Crews and Jensen, 1989

RISK MANAGEMENT AND CAPITAL BUDGETING

Few management decisions are made under conditions where the outcomes associated with each possible course of action are known with certainty. Most major managerial decisions are made under conditions of uncertainty. The frequency of uncertainty in managerial decisions and the risk involved dictate risk analysis be given due consideration in farm and project management decisions.

What is risk? Risk refers to the possibility that some unfavorable event will occur. It is the possibility of loss, injury, or exposure to harm. In aquaculture, risk comes from stock losses. Anything which disrupts the rearing of fish is likely to jeopardize production and marketing of the final product.

The levels of risk vary among species and at different stages of production. The relative lack of knowledge of fish biology in comparison to some land animals makes fish production more risky than the production of food animals. As Secretan (1988) indicates, on a scale of 1 to 100, we know 75 percent of the biology of human beings, and perhaps 50 to 60 percent of the biology of chickens, cows, pigs, and other farm animals, but only about 20 percent of the biology of aquatic species. There are numerous risks involved in the breeding, hatching, and growing of aquatic organisms under intensive management systems.

What sort of risks plague the aquacultural industry? Risks may be classified into two main groups: (1) socio-economic or business risk and (2) physical or pure risks.

RISK TYPES

Socio-Economic/Business

Social Risks

Social aspects of socio-economic risks include changes in tastes, attitudes, or social behavior towards production and consumption of a certain species. The expansion of aquaculture depends on individuals changing their attitudes towards species cultured under intensive closed systems. This may be done through government programs, advertising, and public relations. For example, changes in consumer purchases of catfish have been achieved through advertising and public relations. The growing popularity of catfish may be stifled, however, if "off flavor" problems continue to plague the industry.

Economic Risks

Economic risks such as changes in price of inputs and output, inflation, recession, depression, and other economic conditions which affect national income are primary concerns of commercial fish producers. As demand lags behind supply, producers are concerned that prices will fall. This is presently the case in the U.S. catfish industry. Producers are being warned that they should secure markets before expanding production. Also, the degree of elasticity with respect to supply and demand at both the farm and processor level is a clue to the level of economic risks associated with fish production. Processors facing a more inelastic demand than producers will tend to be less concerned about demand lags. This is one reason that producers are beginning to favor more producer associations or cooperative type marketing organizations.

Marketing Risks

Risks may also result from uncertainty in demand, supply, and prices. When to move the product to market is the age-old nemesis of farmers. Fish farmers are no different. Significant seasonal price level differences exist in many aquaculture product markets. Today, more farmers in colder climates are overwintering fish to try to market them when there is less supply available to consumers. Additionally, new technologies and product forms are being evaluated in an attempt to avoid some of the marketing risks. Smoked fish, as well as dried, frozen, or canned fish are forms used in various markets to reduce the risks associated with marketing time.

Assume that forecasters are overly optimistic in their estimates of prices and consumer demand. This optimism is likely to encourage farmers to intensify production (higher stocking rates) in the short-run and expand production (more ponds) in the long-run. Intensification increases the potential for diseases, problems such as "off-flavor," and other environmental concerns. The fish arrive at the market only to remain unsold because of weak consumer demand resulting from a dislike for the quality of the fish on the shelf, or insufficient income to purchase fish and other market foods.

Longer-term expansion of production means greater amounts of capital and land committed to the aquacultural practice. Because ponds are much easier to build than to remove, these commitments tend to become irreversible, even if prices decline. Once again, market conditions dictate difficulties for the producer.

Production Risks

Many of the marketing risks are also related to *production problems*. Marketing problems may be logistical in nature which may impede production schedules. The timely supply of fingerlings may affect the quantity of foodfish produced at a given time. This may result in grave financial problems for producers. Production risks may also be due to lack of trained manpower to manage the operation. This results in serious constraint or even failure in any aquacultural enterprise.

Other Risks

Other socio-economic risks encountered are *financial* and *political*. *Financial* risks relate to changes in supply of *funds* for production and marketing. Credit restriction and availability often affect the aquacultural industry. Lack of education and understanding of aquacultural production processes among lenders is common in areas where the industry is developing.

Political risks affect not only an enterprise, but the whole sector. Changes in government and governmental *policies* have been known to cause changes in supply and demand of inputs and fish. Governmental regulations may affect all stages and aspects of the industry. Regulations on feed, import of inputs, the introduction of species, and changes in labor laws may greatly influence the industry.

Physical or Pure Risks

Physical risk results from conditions of nature, such as rain, windstorms, clouds, flooding, and drought. Other types of pure risks are plant breakdowns, and failure of safety and other devices. These risks associated with physical or pure risks can be managed to minimize their effects on producers.

RISK MANAGEMENT

Risk is so important in aquaculture that risk management programs should be in place if the industry is to advance. What is risk management? Secretan (1988) defines risk management as the identification, measurement, and economic control of risks that threaten the assets and income of the business venture. Over the years, several strategies have evolved which may be used to reduce risk. Among these are product diversification, alternative marketing periods, insurance, contracting, government programs, third party equity capital, and safety devices.

Diversification

Product diversification in which more than one "cash crop" is produced provides for alternative income streams in the event one fish or species harvest fails. Diversification is one of the oldest methods of risk management. Farmers are known to produce several products in order to spread their risks among many enterprises. For example, though there are many farmers in Mississippi who produce only fish, some of these individuals also invest in the feed and processing industry. In areas where there are many small operations, producers frequently are engaged in several other non-aquacultural enterprises. In some places, polyculture production of rice and fish or crawfish is very popular. Multispecies production is also pervasive in developing economies.

A note should be made that diversification does not always reduce risks. The heightened demands placed on management through diversification may actually induce greater risks because of management stress. However, diversification more often reduces production and physical risk problems.

Continuous or Sequential Marketing

Variability in marketing time may occur when the "crop" is released to the market in a more or less continual stream, or periodically as the market may dictate. Sequential marketing involves the sale of the product at different periods. In the Southeastern U.S., catfish farmers may be forced to sell the product at various periods because of an oversupply of fish during the months of September and October. Jolly and Engle (1988) showed that sequential catfish marketing through processing plants was not profitable unless the price increased substantially to offset holding costs. Cacho et al. (1986) indicated that there was some disagreement among catfish farmers concerning the importance of spreading sales to stabilize prices. In their survey, only 35 percent of the farmers identified this strategy as extremely important in price stabilization.

Formal Insurance

Insurance from either private or governmental sources may help in the event of catastrophic losses, such as fish kills due to weather changes, low dissolved oxygen, or disease. Formal insurance in aquaculture is only beginning to grow in popularity. Private companies are still unwilling to provide insurance to aquacultural firms because of the lack of knowledge of the risks associated with the production of each species. As research in the various areas increases and insurance companies become knowledgeable of the probabilities of failure, they are likely to increase their level of effort in this area.

Future Market or Production Contracts

Contracts provide producers with specific prices and quantities at the end of a specific period. Forward contracting allows a fish farmer to contract with a processor to supply current growing fish stocks or even unstocked fish on some future date at a particular price. This strategy of contracting future supplies is undertaken to reduce the risks associated with market price changes. Cacho et al. (1986) reported that catfish farmers were strongly convinced of the importance of this strategy in reducing risks; about 74 percent of Alabama catfish farmers identified forward contracting as moderately important to extremely important in reducing marketing risks.

Figure 9-1 shows average fluctuations in prices received by fish producers from 1982 to 1988. Obviously, it is more desirable to sell during periods of high prices. Most catfish farmers in the southern United States harvest fish during the months of October and November. Price typically falls sharply during that time period. Yet, even at these low prices, farmers selling fish to processors can still earn a profit. As pointed out earlier, farmers have been moving

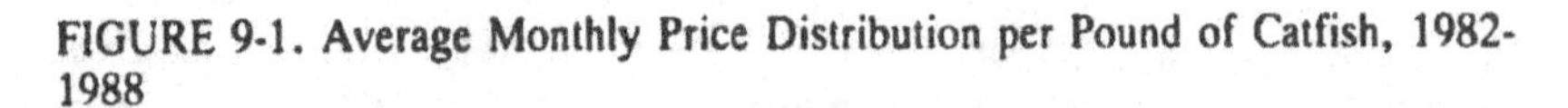

FIGURE 9-1. Average Monthly Price Distribution per Pound of Catfish, 1982-1988

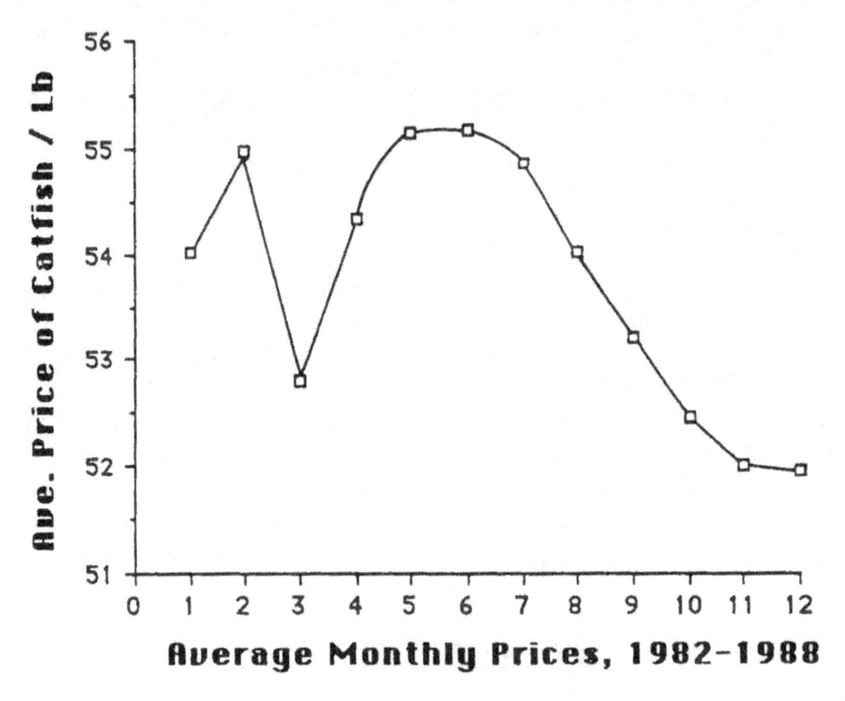

Source: The data were taken from *Aquaculture and Situation Outlook Report*, USDA, September 1989.

toward overwintering fish for sale in March or April when supplies traditionally have been low. Now, however, price declines are occurring in that period. For a producer willing to assume risks of holding fish over the winter, or changing production systems to have fish ready for the market in higher price periods, forward contracting is important. The greater production risks are tempered by guaranteed markets and prices.

The forward contract also may be restrictive on farmer incomes. If market prices exceed contract levels on the due date, the contractor (buyer) gets added benefits. Farmers must sell at contracted prices. An alternative forward contracting approach now evaluated by producers and processors is the production contract.

The production agreement is similar to the forward contract.

Such arrangements are especially popular for highly perishable products such as truck crops, eggs, and fruits. They have been used successfully for a long time by poultry (broiler) farmers. Under this arrangement, producers typically supply the land (ponds) and basic resources (labor and water) to produce the crop. The processor or other buyer provides the fish stocks, feed, chemicals, and other inputs to the producer, then harvests the stocks when they are ready for market. The producer essentially becomes a laborer for the buyer. All risks for production except failure to perform agreed tasks fall on the contracting buyer. To date, fish farmers have resisted formal production contracts. In fact, one farmer organization, Catfish Farmers of America, is urging members and non-members alike to use caution in making such an agreement.

Government Programs

Government programs, especially in the United States, have been used for selected crops such as grains, cotton, peanuts, tobacco, and dairy products. Fish farmers for the most part have had limited opportunity to participate in government programs. As the aquacultural industry grows, it is expected that government programs will be used to reduce risks. In developing economies, there are few direct government programs for aquacultural production. However, there are numerous government programs that may influence producer welfare.

Participation in government programs may be used to finance the operation during difficult times or to purchase needed items to increase production and efficiency in production. Government benefits such as favorable credit terms, low interest rates on pond construction or capital purchases, or even price subsidies are available in some countries. As environmental concerns regarding the quality of aquacultural pond effluent water increases, the likelihood for government programs to assist farmers in reducing risks associated with possible pollution will increase. No matter what the reason given for providing special programs, the reduction in risks is the reason sought. See Chapter 12 for more details on this subject.

Third-Party Equity Capital

The use of outside equity capital allows the fish farmer to transfer some risks to others. Penson et al. (1982) identified three approaches to the use of equity capital: (1) capital leasing, (2) incorporation, and (3) partnership agreements. All three methods are used to some extent in the aquacultural industry. The primary incentive for using outside capital is the high initial cost for a producer. Acquiring the necessary land area, building ponds, and securing an adequate water supply near reliable market outlets may be both difficult and expensive. Additionally, the learning curve for a new producer may be too long for one individual to afford. An added incentive is that having additional capital may mean opportunities for more intensive fish stocking levels, which in turn lead to greater potential profits. Thus, there are circumstances when outside capital is quite beneficial.

The Use of Safety Devices

Many automated systems involving radio signals, computer warnings, and automatic telephone dial-out are presently employed in aquaculture. These systems are complex and may result in losses if they malfunction. However, their use must be considered as a means of reducing farm or processor risks. Secretan (1988) pointed out that risk reducing safety features do not always work. Only limited areas may be monitored. Device effectiveness is tied to operator knowledge and response time relative to the tolerance of a species to the situation of concern. Finally, security devices may engender an unwarranted sense of security which can directly cause losses.

QUANTITATIVE RISK MEASURES

Risk is associated with the probability of undesirable outcomes – the more likely an undesirable outcome, the riskier the decision. The probability of an event is defined as the chance, or odds, that the event will occur. For example, in the production of striped bass, there is little more than a 50 percent chance that the juveniles will survive after stocking, whereas in the production of catfish the chances are 94 percent. If all possible events or outcomes are listed, and if a probability of occurrence is assigned to each event, then the listing is defined as a probability distribution. In stocking catfish fingerlings, the following may be expected:

Event	*Probability of Occurrence*
Juvenile survives	.94
Juvenile dies	.06
	1.00

Notice that the sum of the probabilities add up to 1. The sum of the probabilities of the outcome of an event always add up to 1. Risk may be read from the example above. The risk involved in stocking fingerlings is 6.0 percent.

Take a simple example of two projects to examine how risk may be incorporated in capital budgeting. In project A, the investment is in 1.0 acre of aquaculture and in project B the investment is in 1.0 acre of soybeans. The returns to three scenarios are in Table 9-11.

To evaluate these two projects, probabilities must be assigned to their outcome (see Table 9-12). The expected value of net cash inflows are obtained from use of the formula:

$$E\,(NCI_A) = NCI_1 \cdot P_1 + NCI_2 \cdot P_2 + NCI_3 \cdot P_3,$$

where:

NCI_A = net cash inflow for project A
P_1 = the probability of good weather
P_2 = the probability of average weather
P_3 = the probability of poor weather.

Project A is expected to yield less return than project B if weather conditions are good, but more than project B if average conditions persist. Both projects yield the same if poor conditions prevail. Project A yields higher expected NCI than project B.

Measuring Risk

There is a great deal of controversy as to how risks should be measured. There are several approaches, but one commonly used measure is the "tightness" of the probability distribution. The statistical standard deviation, represented by sigma, σ, is used frequently as the measure of the dispersion of a distribution.

To determine the standard deviation, first calculate the expected value of the NCI. Remember the expected value of NCI is:

$$\text{Expected value} = \overline{NCI} = \sum_{i=1}^{n} (NCI \cdot P_i)$$

NCI_i is the net cash inflow associated with the ith outcome; P_i is the probability of occurrence of that i^{th} outcome; and $\overline{NCI}$ is the expected value of that outcome. The standard deviation is derived from this formula:

$$\sigma = \sqrt{\sum_{i=1}^{n} (NCI_i - \overline{NCI})^2 P_i}$$

The calculation of standard deviation from our two projects A and B may be derived thus:

Deviation of A		Deviation²	Deviation of B		Deviation²
800 − 490 =	310	96,100	1,000 − 430 =	570	324,900
600 − 490 =	110	12,100	400 − 430 =	−30	900
100 − 490 =	−390	152,100	100 − 430 =	−330	108,900

The standard deviation for project A is:

$$\sigma_A = \sqrt{\sigma^2}$$

$$\sigma_A^2 = 96{,}000\,(.2) + 12{,}100\,(.5) + 152{,}100\,(.3)$$
$$= 19{,}220 + 6{,}050 + 45{,}630$$
$$= 70{,}900$$

$$\sigma_A^2 - \sqrt{70900}$$

$$\sigma_A = 266.72$$

The standard deviation for project B is determined the same way.

$$\begin{aligned}\sigma_B^2 &= 324{,}900\,(.2) + 900\,(.5) + 108{,}900\,(.3)\\ &= 64{,}980 + 450 + 32{,}670\\ \sigma_B^2 &= 98{,}100\\ \sigma &= \sqrt{98100} = 313.21\end{aligned}$$

The standard deviation provides information about the distribution of the outcome. It gives information on the "tightness" of the distribution. The larger the standard deviation, the greater the dispersion of that distribution. The measure of absolute dispersion is important, but it is also necessary to measure the relative risk of the cash inflows. This is achieved by the coefficient of variation. The coefficient of variation provides a measure of relative dispersion. The coefficient of variation is obtained by dividing the standard deviation by the expected value. From our previous example, the coefficient of variation (CV) for the two projects are:

$$CV_A = \frac{266.27}{490} = .54 \qquad CV_B = \frac{313.21}{430} = .73$$

The use of the σ to compare riskiness of the two projects could be misleading, especially if the projects were of unequal size. The CV gives the percentage variation of σ over the expected value. Therefore, comparing our two projects, project A has the largest expected value of cash inflow of $490, with a CV of 54 percent, whereas project B has an expected value of cash inflow of $430 and a CV of 73 percent. There is greater variation in return possible with B, as well as a lower NCI.

TABLE 9-11. Payoff Matrix for Two Projects, A and B

State of the weather	Net Cash Inflows	
	Project A	Project B
Good weather	800	1000
Average weather	600	400
Poor weather	100	100

TABLE 9-12. Calculation of Expected Values for Two Projects, A and B

Project	State of the weather	Assigned probability	Outcome if event occurs	Expected value
	Good weather	0.2	800	160
Project A	Average weather	0.5	600	300
	Poor weather	0.3	100	30
	**			
	Good weather	0.2	1,000	200
Project B	Average weather	0.5	400	200
	Poor weather	0.3	100	30

$$E(NCI_A) = 800\,(.2) + 600\,(.5) + 100\,(.3)$$
$$= 160 + 300 + 30$$
$$= 490$$
$$E(NCI_B) = 1000\,(.2) + 400\,(.5) + 100\,(.3)$$
$$= 200 + 200 + 30$$
$$= 430$$

UTILITY THEORY

Another approach used for studying risk is utility theory. In this theory, managers and decision makers are divided into three groups: risk seekers, risk averters, and those who are indifferent to risk. The risk seeker will select the riskier investment. The person who is indifferent to risk is also indifferent to which investment is chosen. The risk averter, given a choice, will select the less risky investment. Underlying the utility approach is that the more profitable the venture, the riskier it is. This theory is based on the marginal utility of money and is somewhat beyond the scope of this book, but theoretically useful.

INCORPORATING RISK INTO CAPITAL BUDGETING

In the evaluation of our previous projects for tilapia and catfish, production risk was not considered in the analysis. A 6 percent mortality was used in the preparation of the catfish enterprise budget, and to some degree this is an element of risk. As was discussed, there are many other factors which may result in fluctuation of net cash inflow over the life of the project. Likewise, there are several approaches for incorporating risk into capital budgeting, but only two will be considered here: the *certainty equivalent* and the *risk-adjusted discount rate*. The *certainty equivalent* approach involves incorporating the utility function of the decision maker into the analysis. The manager substitutes a certain dollar amount believed to be equivalent to the expected, but risky, cash flow offered by the investment in the capital-budgeting analysis. A *set* of risky cash flows is substituted for the original more certain cash flows, between both of which the decision maker is indifferent.

The formula used to obtain the NPV would then be modified by a certainty equivalent coefficient (π_t's) which represents the ratio of the certain outcome to the risky outcome. Thus, the equation for π_t is:

$$\pi_t = \frac{\text{certain net cash inflow}_t}{\text{risky net cash inflow}_t}$$

The πs vary from 0 to 1.

In previous examples, the NPV was found with the formula:

$$NPV = \sum_{t=1}^{N} \frac{P}{(1 + i)^t} - INV$$

Using the certainty equivalent approach, the NPV would be obtained by modifying the formula to:

$$NPV = \sum_{t=1}^{N} \frac{\pi_t P}{(1 + i)^t} - INV$$

where:

π_t = the certainty equivalent coefficient in time period t_1
P = net cash flow
i = the risk-free interest rate
N = the project's expected life
INV = the initial investment.

The *risk-adjusted discount rate* approach is based on the concept that investors demand higher returns for more risky projects. Therefore, the discount rate is adjusted based on the level of risk anticipated in the project. Once a firm determines the appropriate required rate of return for a project with a given level of risk, the net cash inflows are discounted. The normal capital budgeting techniques are applied, except in the case of the IRR. In the IRR case, the required rate (RRR) with which the project's rate of return is compared becomes the risk-adjusted discount rate.

The NPV is now calculated as such:

$$NPV = \sum_{t=1}^{N} \frac{P}{(1 + i^*)^t} - INV$$

where i^* is the risk-adjusted discount rate.

Aquaculture producers generally appear to be risk seekers. With all the potential for failure in aquaculture, the environment is hardly favorable for risk-averse individuals. However, the degree of risk may be estimated and then even the risk seeker has a choice regarding the route chosen.

REFERENCES AND RECOMMENDED READINGS

Barry, P.J., A.J. Hopkin, and C.B. Baker. 1979. *Financial Management in Agriculture*. Fourth Edition. Illinois: The Interstate Printers and Publishers, Inc. 529 pages.

Cacho, O., H. Kinnucan, and S. Sindelar. 1986. *Catfish Farming Risks in Alabama*. Circular 287, Alabama Agricultural Experiment Station, Auburn University, Auburn, Alabama.

Crews, J. and J.W. Jensen. 1989. *Budget Sensitivity Analysis for Alabama Catfish Production*. Alabama Cooperative Extension Service/Auburn University, Alabama 36849, July.

Food and Agriculture Organization of the United Nations. 1983. *Inland Aquaculture Engineering*; Lectures presented at the ADCP Interregional Training Course in Inland Aquaculture Engineering, Budapest, (June, 3 September).

Gittinger, J.P. 1982. *Economic Analysis of Agricultural Projects*. Second Ed. EDI Series in Economic Development, Baltimore: The Johns Hopkins University Press. 505 pages.

Jolly, C.M. and C.R. Engle. 1988. "Effects of Stocking, Harvesting and Marketing Strategies on Profit Maximization in Catfish Farming." *Southern Business and Economic Journal*. 12,1:52-62.

Keown, A.J., D.F. Scott, Jr., J.D. Martin, and J.W. Petty. 1985. *Basic Financial Management*. Third Ed. NJ: Englewood Cliffs. pp. 808.

McGuigan, J.R. and R.C. Moyer. 1989. *Managerial Economics*. St. Paul: West Publishing Company, pp. 653.

Pappas J.L. and E.F. Brigham. 1979. *Managerial Economics*. Third Edition, Illinois: The Dryden Press, pp. 567.

Penson, J.B., D.A. Klinefelter, and D.A. Lins. 1982. *Farm Investment and Financial Analysis*. Englewood Cliffs, NJ: Prentice-Hall.

Secretan, P.A.D. 1988. "Risk Management in Aquaculture, 1988." *Fish Farming International*. 15,2:4-25.

Shang, Y.C. 1981. *Aquaculture Economics: Basic Concepts and Methods of Analysis*. Boulder: Westview Press, pp. 153.

USDA, 1989. *Aquaculture Situation and Outlook Report*. Economic Research Service, AQUA-3, p. 34.

Warren, L.F., M.D. Boehlje, A.G. Nelson, and W.G. Murray. 1988. *Agricultural Finance*. Fifth Edition, Ames: Iowa State University, Press, pp. 468.

Weston, F.J. and E.F. Brigham. 1977. *Managerial Economics*. Sixth Edition, pp. 1030.

Chapter 10

Market Structure and Theory of Price

Studies of the market structure for fish and fish products have concentrated primarily on the marine or seafood industry. The aquacultural market was given little attention until recent years. Limited studies on the aquacultural market structure and conduct have been restricted generally to processed aquacultural products. Thus, it is especially important to examine marketing concepts related to aquaculture with respect to both the fresh and processed products. Such an examination begins with a review of basic marketing concepts.

Market structure is the term given to the *organizational* characteristic of the market which influence the nature of competition and pricing within the market. The structure of a market determines the type of behavior or conduct which prevails in the industry. The market structure for various products encourages certain distinct types of market conduct.

Market conduct refers to the *patterns of competitive behavior* which market participants exhibit in adjusting to the market in which they participate. Four important types of competitive behavior in varying degrees are found in the food industry: perfect competition, oligopoly, monopolistic competition, and monopoly, all of which likely to be found in one segment or another of the aquacultural industry at some time or another.

MARKET PATTERNS

Pure Competition

The competitive model is based on assumptions that are very stringent and unlikely to be satisfied in any real world market. Industries that do exhibit many of the following characteristics are said to be perfectly competitive.

According to theory, under pure competition, demand for most commodities comes from a large number of buyers, who act quite independently from one another, rarely making any conscious effort to influence price. Supply is also provided by many small producers whose individual production is not sufficient to influence market price. Therefore, the essence of the concept is that the market is entirely impersonal. That is, the bargaining strength of buyers and sellers is weak.

The perfect competition model possesses the following characteristics: (1) each economic agent acts as if prices are given, that is, each acts as a price-taker; (2) the product is homogeneous; (3) there is free mobility of all resources, including free entry and exit of firms; and (4) all economic agents possess complete and perfect knowledge of the market. Small scale artisinal fisheries in many developing countries are considered purely competitive. Fishermen use open access resources and there are numerous operators. Aquacultural markets for tilapia in many developing countries such as Haiti and Rwanda may be classed as pure competition since they meet practically all the conditions. Fish production usually is done in family ponds requiring relatively small investment. Most of the investment is in terms of family labor for pond construction. Fish produced are for family consumption. Any surplus is sold or traded to other non-producing families.

Conditions for Pure and Perfect Competition

Price-taking buyers and sellers. In this economic model, the fish producers or marketing agents, do not take into account the effect of their behavior on price when making a consumption or production decision. The firms usually are small, relative to the market as a whole, and cannot exert a perceptible influence on price. For example, aquacultural enterprises in Rwanda typically range from .01 hectare (.025 acre) to 9.0 hectares (22.5 acres). The majority are less than .02 hectare (.05 acre). Purchasers buy very small quantities of fish at any one time. None of the participants exert any influence on the market. In this and other similar markets, base prices are established at places other than the specific market site. Local producers and buyers accept those prices with minor exceptions. Any price variations account for differences in transportation, quality, etc.

Homogeneous product. The products must be virtually identical. The product of one seller is the same as that of another. This ensures that buyers are indifferent as to the firm from which to purchase. If buying catfish, the purchaser evaluates the catfish from producer A as being the same as from producer B. Then the product can be considered homogenous. In Rwanda and Haiti, almost all producers grow tilapia to a given size, and most produce and sell *Tilapia nilotica*. Hence, the product is virtually identical among producers as well as in buyer's minds.

Free mobility of resources. All resources must be perfectly mobile. That is, each resource may be moved in and out of the market readily in response to monetary signals. This is only so, however, in the long-run since time is required to change fish production facilities. Land use patterns and regulations may also affect resource mobility. In some countries, land is the most important resource for fish production. If the producers own their land, they may produce whatever and whenever desired. If land must be leased, resource mobility may be restricted.

Perfect knowledge. Consumers, producers, and resource owners must possess perfect knowledge if a market is to be perfectly competitive. If consumers are not fully cognizant of prices, they might buy at higher prices when lower ones are available. Producers must know their costs as well as price in order to attain the most profitable rate of output. Since few cases occur where there is absolute knowledge of the market by both producers and consumers, it is sufficient to say that perfect competition *may* exist where there is a high level of knowledge about competing products, production processes, and consumer needs.

The Fish Farmer Demand Curve

The demand curve for a particular fish farmer is depicted in Figure 10-1. Since there may be many firms operating in the market and selling the same product, the demand curve is *highly*, though

FIGURE 10-1. A Hypothetical Demand Curve Faced by the Individual Fish Farmer

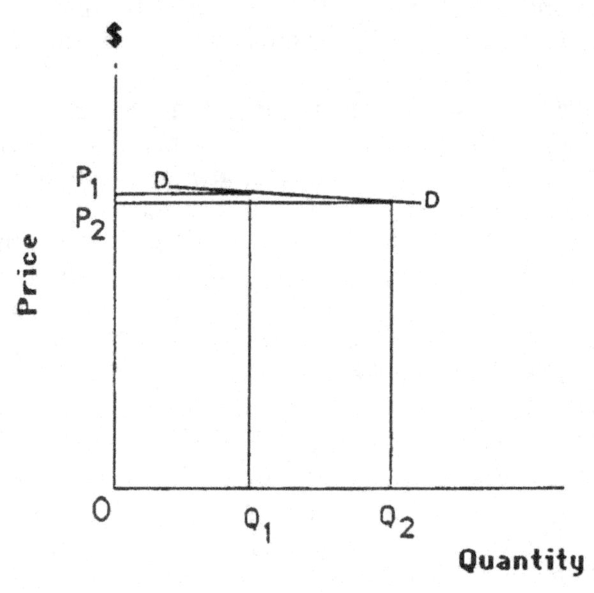

not *perfectly*, elastic. The curve is portrayed as being almost parallel to the X-axis since the action of the buyer cannot affect price. If the market price changes only slightly, from P_1 to P_2, quantity is likely to change from Q_1 to Q_2. However, if the individual fish farmer is a price taker and must accept the going market price, then demand would be perfectly elastic, either P_1 or P_2, depending on supply conditions.

Pricing Under Pure Competition

As discussed in Chapter 2, under pure competition market price is determined by the interaction of supply and demand. With so many buyers and sellers having knowledge of the market conditions, no one individual can influence price. Every buyer will be trying to buy at the lowest price while every seller will be trying to sell at the highest price. The market price is called the equilibrium price for the whole market (see Figure 10-2). Recall from earlier discussions that the relevant portion of the marginal cost curve is

FIGURE 10-2. Market Demand and Supply Curves for a Purely Competitive Market

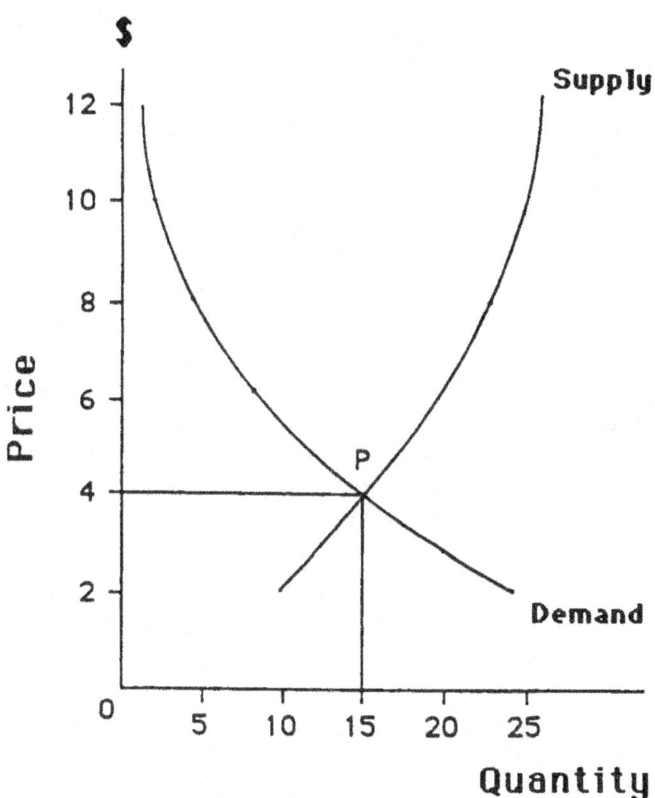

also the short-run supply curve. Also recall that if a producer is a price taker, prices are dictated by the total market. Price fluctuation is not common. Under such conditions, the price of the product which is its average revenue (AR) is also its marginal revenue (MR).

Long-Run Profit Maximization

Each fish farmer will be continuously adjusting output to maximize profit. Profit is the difference between total revenue (TR) and total cost (TC). In the perfectly competitive market situation, profit is maximized when marginal revenue equals marginal cost. Marginal revenue also equals price since all producers are price takers. The farmer is maximizing profit at the point where marginal cost (supply) equals price (MR). Using Figure 10-3, in the short-run situation, farmer A will be making a profit. In a competitive market where there is free mobility of resources, equilibrium will be where MC = MR and also where MC and MR equals long-run average total cost (LATC).

If above-average profits are being made, there will be more fish farmers entering the industry. As farmers enter the industry, they increase the supply of the product and thereby reduce market price. For a farmer to be in equilibrium in the long-run, price must be equal to short- and long-run marginal cost, and the farmer must be satisfied with the existing plant scale and must not be making economic profit. Economic profit is revenue in excess of all costs including normal returns to fixed resources such as labor, land, and management. The full long-run equilibrium for the farmer operating under purely competitive market situations must be where SMC

FIGURE 10-3. Use of Long-Run Marginal Cost Curve

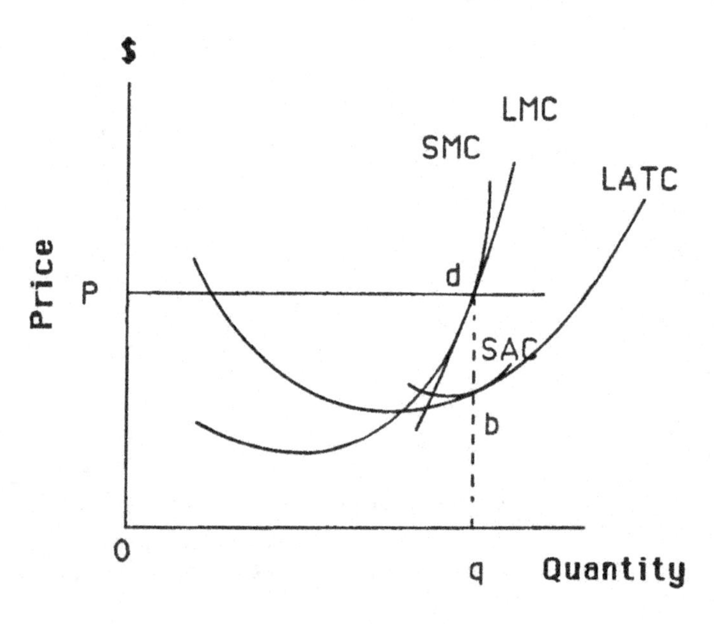

equals LMC equals SAC and equals LATC (see Figure 10-4). This will be caused by a change in price due to changes in supply, and also by changes in the cost of the factors of production due to change in demand for these factors.

Monopoly

A monopoly is said to exist if there is one and only one seller in a well-defined market. Perfect competition is the opposite of monopoly. There are no direct competitors or rivals, nor are there close substitutes for the product. However, the policies of a monopolist may be constrained because of the threat of competition from others who may envy the monopolist's position. Producers of other goods, which may be good substitutes, and producers of other non-related goods, which compete for the same consumer spending, are primary competitors for monopolists.

FIGURE 10-4. Long-Run Equilibrium of a Purely Competitive Market

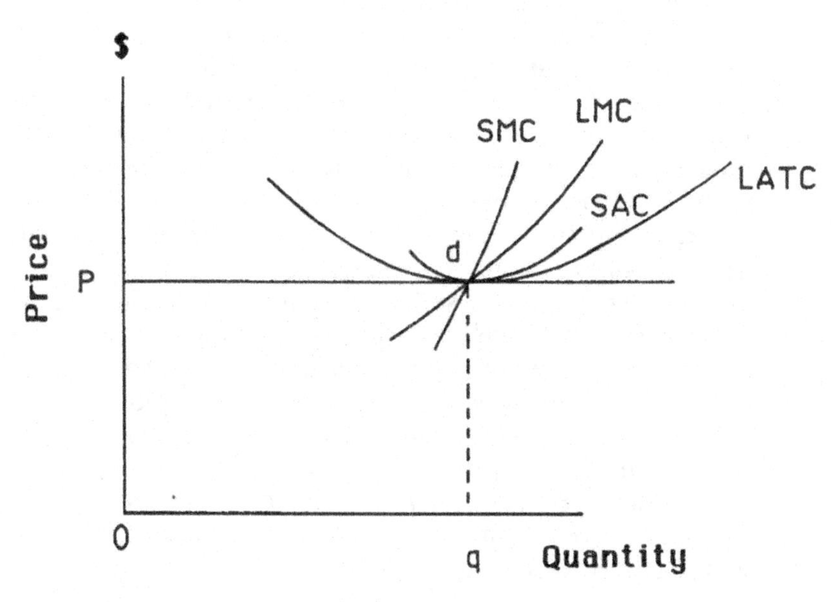

Bases for Monopoly

1. There is only one producer. The producer determines the amount to be produced and the price at which it should be sold.
2. Close substitutes are rare or unavailable.
3. There is no freedom for new firms to begin producing the product. Through control of the raw materials market, the existing firm may prevent new entries. New firms may be restricted from market entry through regulations and patents.

If a monopolist makes above-average profit, rival entrepreneurs will try to enter the industry. For monopoly to survive, there must be certain conditions including:

1. *Barriers to Entry*. There are laws and legal impediments such as patents which prevent free entry and exit. However, market constraints through competitive pricing also are used to prevent entry by competitors.
2. *Economies of Scale*. The nature of the long-run average cost curve of the industry may permit only a single firm to operate. This firm then becomes the industry. If one firm is in control of marketing outlets, the firm may become vertically integrated and then control production and cost conditions.
3. *Ownership of Vital or Superior Resources*. If a firm owns all the known supply of raw material necessary to make a given product, new rivals cannot effectively enter the market. For example, if there is only one fish farmer who has knowledge on producing fingerlings, the market may be monopolized. This usually is temporary because it does not take long before other producers gain the skill and begin producing fingerlings.
4. *Imperfect Capital Market*. Large sums of money typically are required to initiate aquacultural projects. Banks and financial markets are not always willing to finance such projects, and only one or a few individuals may have the financial status to begin them. This may occur, for example, in the feed production business. However, farmers may combine resources to produce their own feed, thus reducing their dependence on one source of feed supply.

Demand and Marginal Revenue Curves for a Monopolist

Since the monopolist represents the industry, the market demand curve is the monopolist's demand curve. The demand curve is negatively sloped. At lower prices, consumers typically buy more. The marginal revenue curve, which is derived from the total revenue curve, is also downward sloping and not horizontal, as in the case of the purely competitive firm. A change in the market price for the monopolist will affect the quantity sold. Since the monopolist must lower the price to sell an additional quantity, marginal revenue is not the market price.

Marginal revenue for the monopolist is the additional revenue attributable to the addition of one unit of output to sales per time period. After the first unit sold, the marginal revenue is less than price (see Figure 10-5).

Relationship Between Demand and Marginal Revenue

The relationship between total revenue, marginal revenue, and the demand curve was explained in Chapter 3. In evaluating these concepts for the monopolist, it must be remembered that the monopolist represents the entire market. When demand is linear, MR is a negatively sloped straight line that lies exactly halfway between the vertical axis and the demand curve. This relationship results because MR is the first derivative (slope) of the demand (AR) function, and the curves reflect that condition.

A monopolist will maximize profit (minimize loss) when marginal revenue equals marginal cost. Whether a profit or loss is made depends upon the relation between price and average total cost (see Figure 10-6; profit is the area PBDC).

Few monopolists exist in the aquacultural subsector. In the early stages of aquacultural development, a single farmer or the government frequently acts as a monopolist in supplying eggs or fingerlings. The nature of the industry cost conditions sometimes does not permit many farmers to operate simultaneously. If two or more

FIGURE 10-5. Total Revenue Demand and Marginal Revenue for a Monopolist

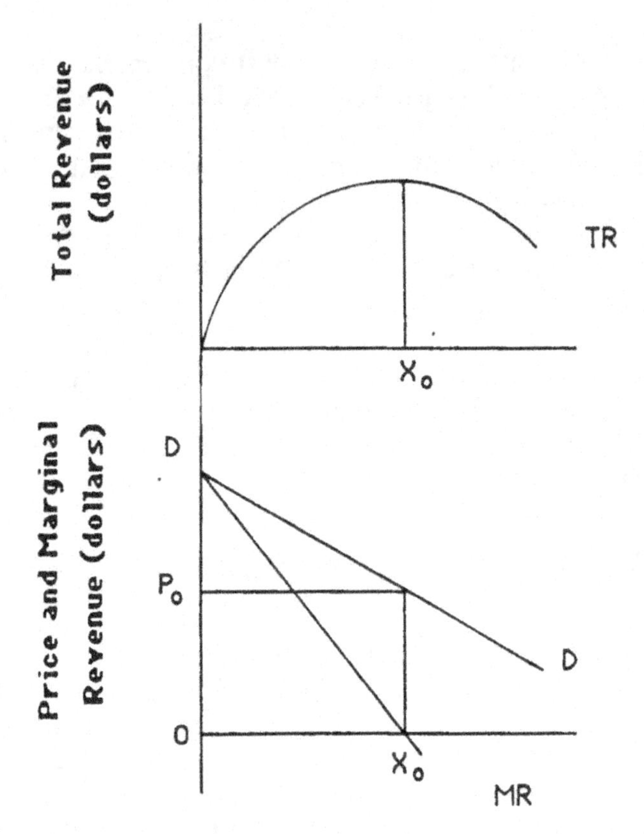

farmers enter the business of fingerling production, the operation may be unprofitable. Fingerling production in the early stages of the industry is rather capital-intensive and has a skilled labor requirement which is more than the beginner farmers can provide. Such a situation is called a *natural monopoly*. As farmers increase their operations and learn the technique of fingerling production, they do not depend on a single hatchery, but often choose to produce their own fingerlings. Consequently, the prevalence of monopolies in the

FIGURE 10-6. Short-Run Equilibrium Under Monopoly

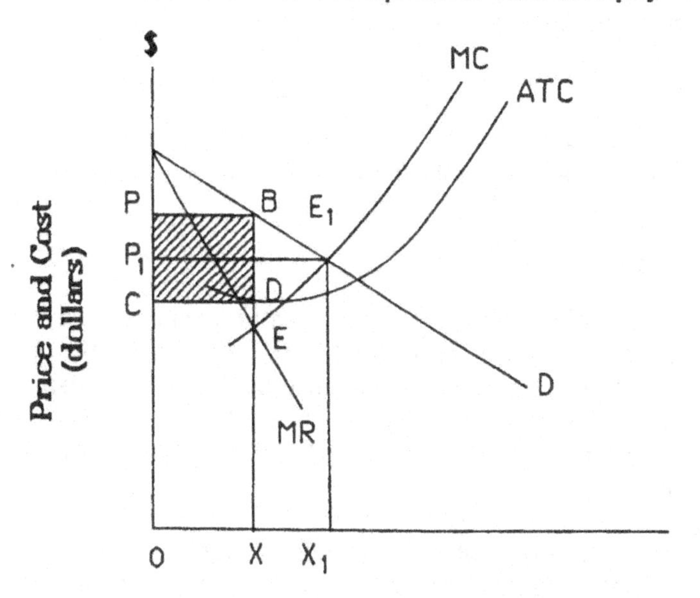

sub-market diminishes and more normal competitive conditions emerge.

The Effects of Monopoly

The market structure under monopoly is always a concern to administrators and governments. In many countries, monopolies are banned and considered bad for the industry and society. This is especially so in the developing stages of a new industry. Monopolies normally are associated with many social evils such as higher prices, lower output, and a loss in consumer and societal welfare. In Figure 10-6, under a purely competitive situation, X_1 of fish would be produced and sold at price P_1. Consumers would be receiving more fish at a lower price. There are conditions, however, in which monopolies may be preferred, at least in the short-run: for example, public utilities. Because of the high capital costs and extensive distribution system, one firm may be preferred over many. Frequently, governments will legalize public utility monopolies, and then regulate them with respect to price or output, or both.

Oligopoly

Oligopoly is said to exist when there are only a few sellers in the market, but the number is not large and the sellers recognize their mutual interdependence.

Characteristics of Oligopoly

1. A small number of large firms each producing (or buying) a large enough portion of the total to affect prices and the pricing policy of the others. There may also be one or two small firms coexisting with a few large firms. However, the small and large firms usually operate in different sub-markets or market sectors.
2. The product produced has no close substitute. Through advertising and promotion the firm will attempt to convince consumers there are differences among the products of competing firms.
3. There is limited entry into the industry by new firms. Entry barriers are similar to those of the monopolists.
4. There is little collusion among firms, but recognition of mutual dependence tends to reduce overt price competition.

Demand Curve of an Oligopolistic Firm

Demand curves depicted in Figure 10-7 cross at the prevailing market price. In this model it is assumed that only two firms exist. The difference between the two demand curves is the price elasticity of demand. In an oligopolistic market structure, if one firm changes price, other firms will react by changing prices. Under oligopoly, the share of a particular market that may be obtained is often more important than profit margins on particular products. Consequently, the demand curve for the initial firm shifts position so that instead of moving along a single demand curve as it changes price, the firm moves to an entirely new demand curve (Figure 10-7).

The demand curve D_1 represents the area of demand where the

FIGURE 10-7. Shift of the Demand Curve by Firms Operating in an Oligopolistic Market

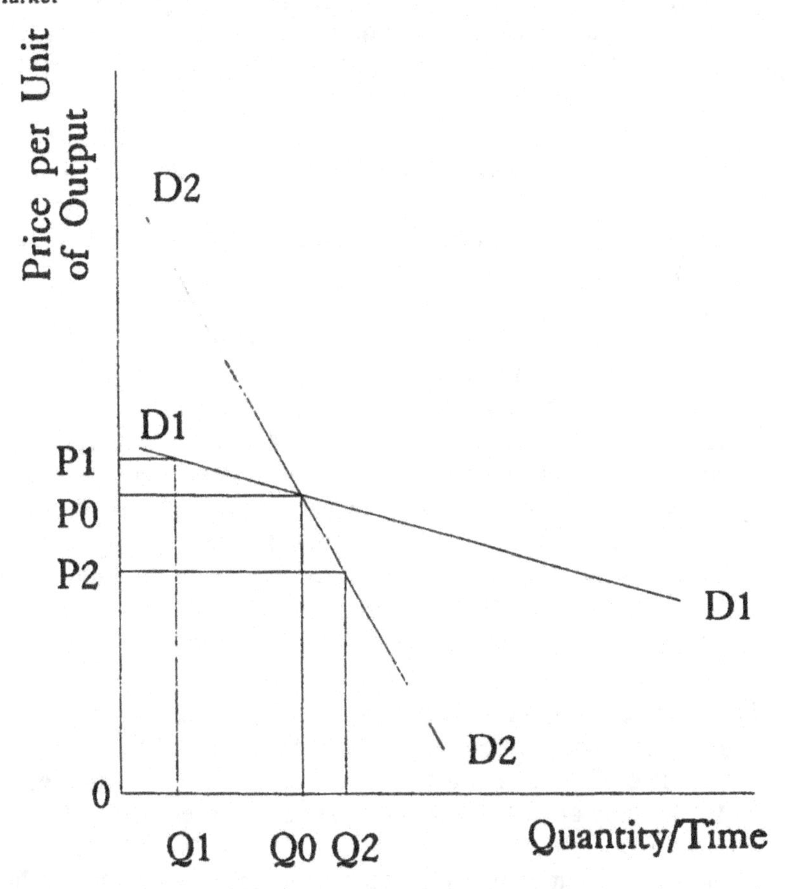

second firm will not follow. If the firm with demand curve D_1 raises price to P_1, the other firm will sell larger quantities at a lower price. All firms will prefer to sell quantity Q_0 at price P_0.

Price Leadership

A less formal, but nonetheless effective means of reducing oligopolistic uncertainty is through price leadership. Price leadership results when one firm establishes itself as the industry leader and spokesperson, and all other firms in the industry accept its pricing policy. This leadership may be the result of the recognized ability of the leader to forecast market conditions accurately and to establish a price that produces satisfactory profits for all firms in the industry.

MONOPOLISTIC COMPETITION

Monopolistic competition has characteristics of both perfect competition and monopoly, including:

1. Many small producers, each having a small market share with little influence on price.
2. Output is not highly standardized, but the output for each is a fairly good substitute for the output of others; the producers, through promotion and advertising, attempt to convince consumers there is a real difference in their product and all others.
3. Little or no collusion among firms, each generally ignores the existence of other firms. There usually is fierce competition for market shares.
4. There are few barriers to entry into the market by other firms.

Examples of monopolistic competition are found in the production of aquacultural equipment, such as aerators. All firms producing aerators design them slightly differently and each is guarded by patents, but they are all reasonably good substitutes with respect to function.

In the equipment industry, each firm tries to make its product unique or different from that of every firm. In a sense, each firm is a "little monopoly." But the monopoly is of very limited power because from the fish producer's perspective, the products of competitors are close, but not perfect, substitutes.

The demand curve faced by each firm is no longer the horizontal, perfectly elastic one that faces the firm under pure competition. Because of product differentiation, a demand curve has a negative slope, but it is still highly elastic because of closeness of substitutes (see Figure 10-8). Price under such conditions will also be similar among firms, not because of fear of retaliation of rival firms, but

FIGURE 10-8. An Elastic Demand Curve for a Monopolistic Competitive Firm

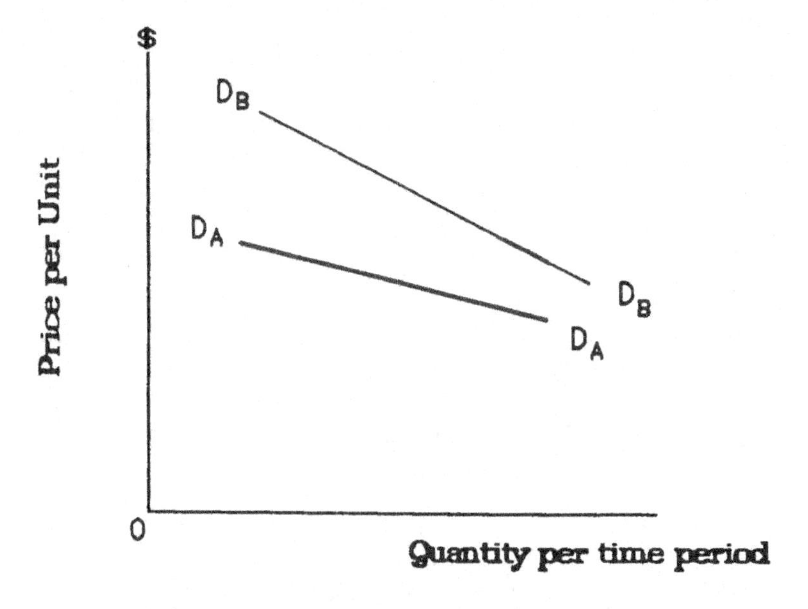

because of the likely loss of consumer markets if prices are out of line.

Price Output Decisions Under Monopolistic Competition

The equilibrium situation for a representative firm producing in a monopolistically competitive market is shown in Figure 10-9. With the demand curve D_1 and the related marginal revenue curve MR_1, the optimum output Q_1, is found at the point where MR = MC. But here, price P_1 is just equal to average cost per unit ATC, so profit is zero.

Now suppose that consumer incomes rise, causing demand to shift to the right, say D_2 and MR_2. The optimal output would be Q_2. The additional profit under that condition would cause firms to enter the industry and lower profits back toward the zero level.

Large group, long-run equilibrium under price competition in a monopolistically competitive product group is attained when the anticipated demand is tangent to the long-run unit cost curve. If

FIGURE 10-9. Price and Output Relationship for a Monopolistic Competitive Firm

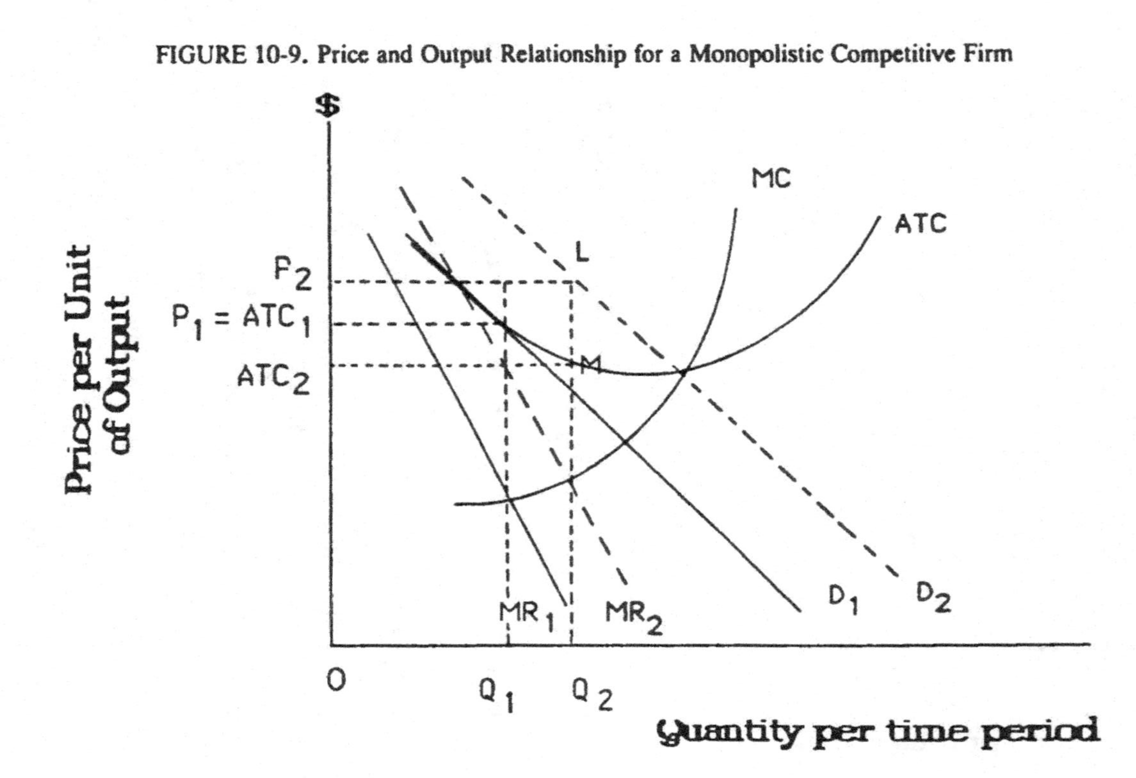

there is no price competition (collusion) but there is free entry, equilibrium occurs where actual demand is tangent to LAC (Figure 10-10). This equilibrium has the short-run characteristic that no firm has an incentive to alter its price or output since MR = MC at Q_e.

Cartels and Mergers

A cartel is an agreement among independent sellers to act together with regard to their marketing decisions. Although these sellers maintain their independence as separate business entities, they make decisions with the welfare of the group as a whole in mind rather than engaging in rival behavior with each other.

If a cartel has absolute control over all the firms in the industry, it can operate as a monopoly. To illustrate, consider the situation shown in Figure 10-11. The marginal cost curves of each firm are summed horizontally to arrive at an industry marginal cost curve. Equating the cartel's total marginal cost with the industry marginal revenue curve determines the profit-maximizing output and price to be charged.

Once the profit-maximizing price/output level has been determined, each individual firm finds its output by equating its own marginal cost to the previously determined industry profit-maximizing marginal cost level.

A merger is the formal joining together of two or more separate business firms into a single business entity. A horizontal merger involves two or more firms that sell the same product in the same market, while a vertical merger involves two or more firms that operate at different stages of the creation and sale of a product.

FIGURE 10-10. Long-Run Equilibrium for a Monopolistic Competitive Market

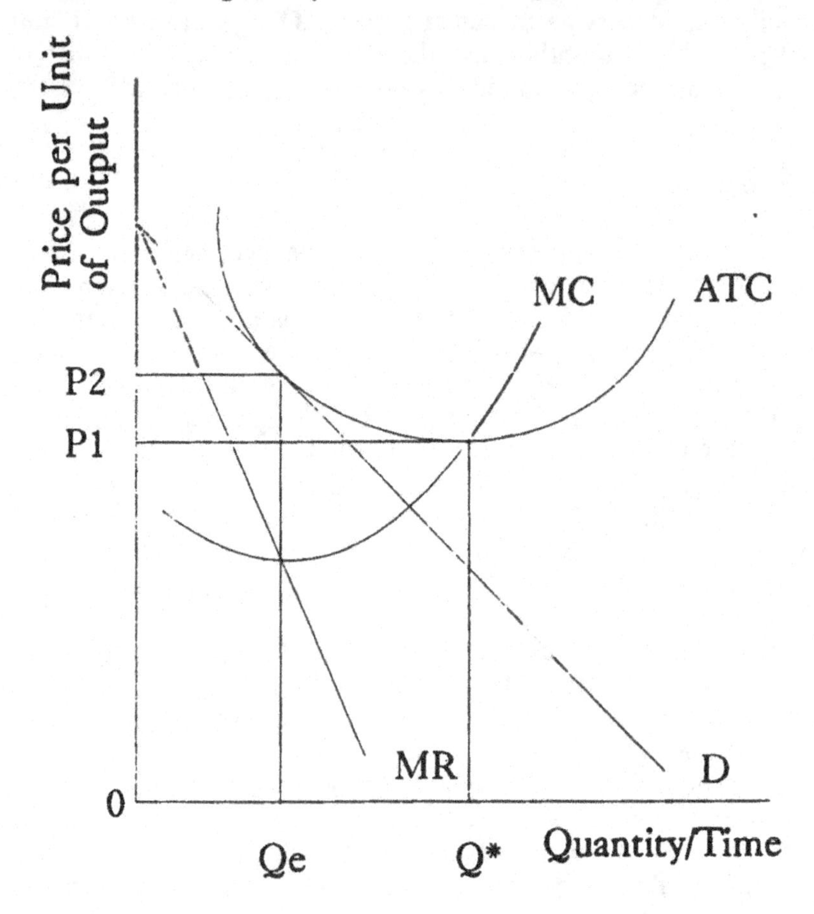

FIGURE 10-11. Price-Output Determination for a Cartel

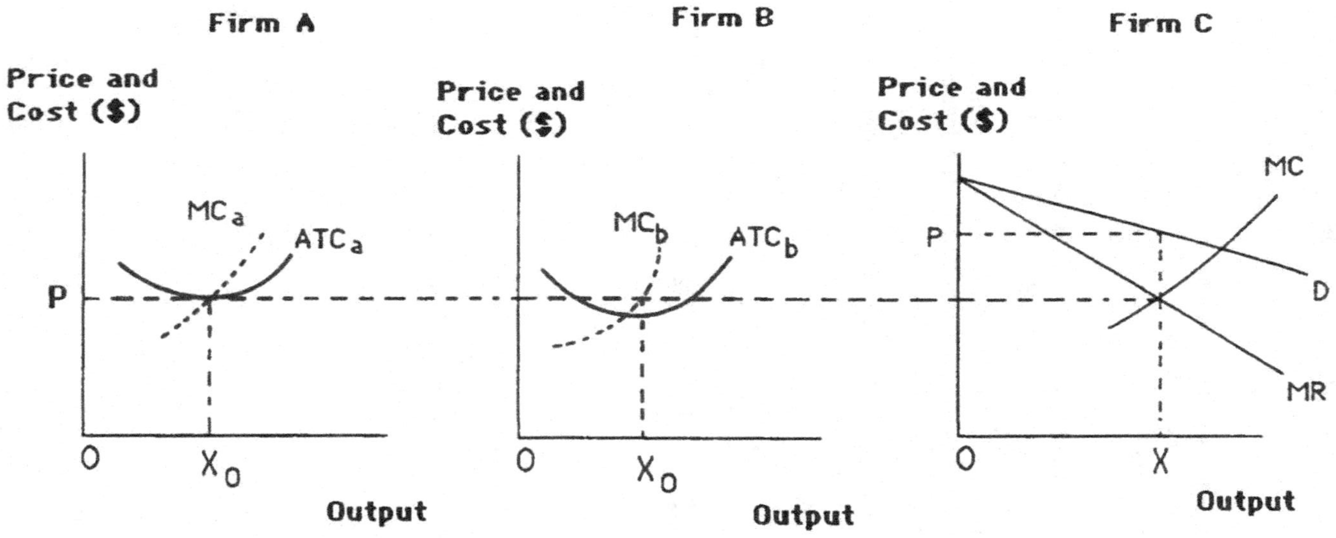

REFERENCES AND RECOMMENDED READINGS

Browning, E.K. and J.M. Browning. 1983. *Micro-economic Theory and Applications*, Second Ed. Boston: Little Brown, Co. pp. 583.

Dellenbarger, L., E.J. Luzar, and A.R. Schupp. 1988. "Household Demand for Catfish in Louisiana." *Agribusiness: An International Journal*. 4(1988):493-501.

Fuller, M.J. and J.G. Dillard. 1984. *Cost-Size Relationships in the Processing of Farm-Raised Catfish in the Delta of Mississippi*. Bulletin 930, Mississippi State University, December.

Kinnucan, H. and G. Sullivan. 1986. "Monopsonistic Food Processing and Farm Prices: The Case of West Alabama Catfish Industry." *Southern Journal of Agricultural Economics*. 18:15-24.

Miller, J.D., J.R. Connor, and J.E. Waldrop. 1981. *Survey of Commercial Catfish Processor: Structural and Operational Characteristics and Procurement and Marketing Practices*, Agr. Econ. Res. Report, No. 130; Mississippi State University.

Raulercon, R. and W. Trotter. 1973. *Demand for Farm-Raised Channel Catfish in Supermarkets: Analysis of Selected Market*. USDA, Economic Research Service. Marketing Research Report, No. 933.

Simon, J.L. 1975. *Applied Managerial Economics*, Englewood Cliffs, New Jersey: Prentice-Hall, Inc. pp. 475.

Chapter 11

Marketing

Aquaculture remains a relatively new industry worldwide. In recent years, the industry has experienced rapid expansion in both developed and developing areas of the world. The development of the industry, however, has been plagued by marketing problems. One would expect that since it is a relatively new industry, management would use a modern approach of matching production to market needs. The catfish industry in the United States, for example, was well advanced into production and processing before producers and administrators gave serious attention to market research and market data collection. In the early years, demand for aquacultural products was so great that producers and processors had no difficulty selling all they had or could produce. Now the situation is changing, and market development is essential for the industry's survival. Data collection on production, processing, and prices on various aquacultural species began only in the 1970s.

Examples of the dilemma in aquacultural marketing are evident in the salmon and shrimp industries. The market for cultured salmon increased rapidly in the 1980s. Norway produced 27,200 metric (27,635 U.S.) tons of salmon in 1985 and planned to increase production to 80,000 metric (81,280 U.S.) tons by 1990. The demand for cultured salmon in the United States has been also rising. In 1984, the United States replaced France as the leading importer of Norwegian cultured Atlantic salmon. Consequently, Pacific salmon fishermen are concerned about the impacts of cultured salmon imports on Pacific salmon prices. Alaskan fishermen, fearing loss of market shares, even oppose permitting salmon farming in Alaska.

As the demand for cultured salmon increases, market information must be available to provide consumers and producers with knowledge on prices and quantities so that (1) market efficiency may increase; and (2) U.S. producers may participate in the production of cultured salmon or substitutes for this high quality product.

The shrimp market also developed rapidly over the past 10 years. The United States took the lead in demand for shrimp imports during the period 1984-1986. Imports rose to meet domestic demand and a total of 400 million pounds were imported, mostly from Mexico. As the demand for this product increased, U.S. producers who found it difficult to produce shrimp domestically, set up farms in neighboring countries. The result is evolvement of a highly complex marketing system that requires much study to comprehend. And, because of the market complexity, many promising production or processing units have failed.

Knowledge of the marketing process is required for locating markets for new and established products, for price determination, and for setting quality standards for aquacultural products. Today the Catfish Institute of America is aiming to ensure quality so as to increase sales by making their product, catfish, a household word. To ensure quality, the industry has begun an inspection program in participating processing plants. A number of advertising campaigns have also been conducted with the aim of increasing total sales. Fish farmers might not want to devote personal time to marketing activities, but they have little choice if they wish to succeed in the long-run. Marketing will become an area of specialization in high demand as consumption of aquacultural products increases in the U.S. and around the world.

WHAT IS MARKETING?

There is no generally agreed upon definition of marketing, although many working definitions are given by professionals. Those with an agricultural orientation usually stress the functional aspects of activities involved in movement of farm-produced raw material from the farm to the ultimate consumer. Functions such as production, hauling, processing, storage, wholesaling, and retailing are emphasized. Efficiency is a major consideration from the standpoints of location, transportation, capacity, and seasonality. Others, more interested in market organization and efficiency in management, favor a market structure approach. This approach to marketing includes the system of markets and related institutions which organize the economic activity of the food and fiber sector of the economy.

A structural approach permits coordination of economic activity through institutional arrangements and is not entirely dependent upon market prices to perform their traditional function. Rather, there is a recognition that structural arrangements in most economic sectors are far removed from the perfect market concepts. Consequently, government may have a legitimate role in taking steps which would be helpful in permitting the market to perform satisfactorily.

Marketing Defined

Since concern in this book is with aquacultural products, the functional approach to marketing will be stressed. A definition for marketing under this approach is:

> Marketing of aquacultural products is the performance of all business activities involved in the flow of aquacultural products and services from the point of initial aquacultural production until they are in the hands of consumers. Marketing is the name given to the management process responsible for finding out what customers need and supplying them as efficiently and profitably as possible. Marketing begins on the farm and ends with the satisfied consumer.

Fish marketing is neither a mechanical nor an automatic operation. It can be a complex process where the product is changed in form, such as from fish to fish cakes, and the product undergoes many buying and selling transactions before reaching the ultimate consumer.

Marketing has a central role in the management of a fish farm. Decisions made on any aspect of fish farm management which affect customers and producers are part of marketing and must be considered from the marketing point of view. Even decisions which may seem purely technical may have a marketing element.

A market may also be described as an arena for organizing and facilitating business activities and for answering the following basic economic questions:

1. What to produce? This includes the species, the size, the quality, and the price the consumer is willing to pay for the product.
2. How much to produce? How much of each species, and each size and quality to produce at different periods?
3. How to distribute production? What mechanisms will be used for distribution? What type of networks will be used?

A market may be defined by:

1. a location – the New York fish market;
2. a product – the shrimp market;
3. a time – September-October catfish market;
4. a level – retail market.

Marketing is productive. It creates utility, that is, marketing makes goods and services useful. The utilities created are:

1. place – the transfer of the fish from farm gate to supermarket;
2. time – the overwintering of live fish or storage of processed fish products;
3. form – the transformation of fish into fish steaks; and
4. possession – the consignment of fish from wholesale to retailer.

The marketing process is one of movements. It is a series of actions and events that take place in some sequence. Thus, some form of coordination for this series of events and activities is necessary if goods and services are to move in an orderly fashion from fish producers to fish consumers.

MARKETING FUNCTIONS

A marketing function may be defined as a major specialized activity in accomplishing the marketing process. The marketing function may be described as follows:

A. Exchange function
   1. buying (assembling) of various species of fish or fish products
   2. selling

B. Physical function
   3. storage of the fish product
   4. transportation
   5. handling and Processing

C. Facilitating
   6. standardization
   7. financing
   8. risk bearing
   9. market intelligence

A. The *exchange functions* are those activities involved in the transfer of title of the fish. They represent the point at which a study of price determination enters into the study of marketing.
   1. The buying function is largely one of seeking the sources of supply, assembling fish products, and activities associated with purchase. This may be assembling fish and the activities associated with the purchase of fish from small producers by a processor or middleman, or fish received and assembled from producers or captured from the wild.
   2. The selling function involves a series of actions called merchandising. Most of the physical arrangements or display of goods are grouped here. Advertising and other promotional efforts to influence or create demand for fish are also part of the selling function. This involves packaging, displaying, and advertising.

B. The *physical functions* are those activities that involve handling, movement, and physical change of the actual commodity itself. They are involved in solving the problems, when, what, and where in marketing.

3. The storage function is primarily concerned with making fish available at the desired time. It may be done in the pond, holding tanks, refrigerators, or in storehouses.
4. The transportation function is temporarily concerned with making fish available at the proper place. Adequate performance of this function requires the weighing of alternative routes and means of transportation as they might affect transportation costs. In developed countries, the process is very sophisticated. Fish may be transported live in tanks, or dressed in large refrigerated trucks. In developing economies, fish may be transported in ice boxes or plastic bags filled with water, or dried prior to shipment.
5. The handling and processing function involves all those basic activities such as freezing, or changing live catfish into fish nuggets and fish cakes that change the product. The simple act of drying fish is also a processing function. The form of the product is changed.
   a. Freezing, at present is the only method that can preserve fresh fish characteristics during long storage. In addition, if applied on a large scale, it offers several advantages common to industrialized processing:
      - consistent quality;
      - product variety;
      - possibility of stabilizing supply and price;
      - hygienic packaging and distribution;
      - standardization of product type;
      - extension of range of retail outlets;
      - creation of incentive for manufacturers to use modern advertising;
      - creation of new marketing strategies and extension of market shares; and
      - reduction of market risks.
   b. Drying comprises a variety of processing techniques aimed at the preservation of fish, by dehydration or addition of a chemical substance, or both. Additives and chemical changes effected during curing may impart either a desired or undesired flavor to the product. Drying either alone or in combination with salting or smoking, or in

combination with both, still accounts for the largest share of the fish used for processing for human consumption. In tropical countries with an abundance of sunshine, fish drying is prevalent. However, losses may occur if sun drying is done improperly.

c. Salting is the process in which salt is the chief preservative. Markets for salted fishery products have been on the decline in developed countries as a result of changes in dietary habits and health consciousness. Salt acts as a preservative by extracting water from the raw material. Quick penetration of the salt into the tissues of the fish is desirable to protect quality.

d. Smoking is the process in which drying and the addition of chemicals (usually salt) are combined with smoke from burning wood for the preservation of fish. Proper smoking may increase fish value.

e. Canning for preservation of fish requires that a number of basic conditions be met:
   - Fish species suitable for canning should be available in sufficient quantity and quality at regular intervals and at a price less than that of the market price for "grade A" fish.
   - Products must be acceptable to the consumer. That is, canned fish products must be part of the cultural norm of the population.
   - Access to adequate communication and energy sources must be present.
   - There must be a relatively large supply of unskilled labor that can be rapidly trained to perform the bulk of the canning operations not requiring special skills, and a small number of highly skilled personnel for supervisory and technologically specialized jobs.
   - Canning materials must be readily available at reasonable prices.

f. In the fish-reduction industry, reduction operations are based on the utilization of fishery resources which can be used more profitably for industrial than for human consumption purposes. Fish meal and fish oil are ex-products of fish reduction. Other techniques such as radiation and freeze-drying are now at the experimental stages. Some are already in use at the market level.

C. The *facilitating functions* are those that make possible the smooth performance of the exchange and physical functions:

6. The standardization function is the establishment and maintenance of uniform measurements and grades.
7. The financing function is the use of money to carry on the various aspects of marketing.
8. The risk bearing function is the acceptance of the possibility of loss in the marketing of the fishery product.
9. The market intelligence function is the job of collecting, interpreting, and disseminating the large variety of data necessary for the smooth operation of the marketing processes.

MARKETING CHANNEL

Some form of institutional structure must be in place for the movement and exchange of goods to take place. When an aquacultural product is transferred from producer to consumer, and especially if the product is marketed through a structured channel, relatively good records are maintained. As the product moves from producer to consumer there is a flow of information generated. This flow of information is essential to any marketing channel. If there is mutual agreement over the conditions of the sale, a physical transaction flow involving the product and payment will occur. In Figure 11-1, a demand for fish by the consumer generates a flow of information. The information flow is from consumer to producer, each sending messages of price, quantity, quality, and availability through a channel connected by intermediaries.

The marketing channel is the network linking the producer to the final consumer. The channel may be simple, comprising only a few organizations such as a single fish producer and a single consumer, or complex with numerous intermediaries between producer and final consumer. The marketing channel for all species follow similar movements as the fish pass from producer to consumer.

Some fish farmers are able to sell directly to the final consumers for several reasons. The quantity of fish might be so small that the

FIGURE 11-1. Market Information Flows

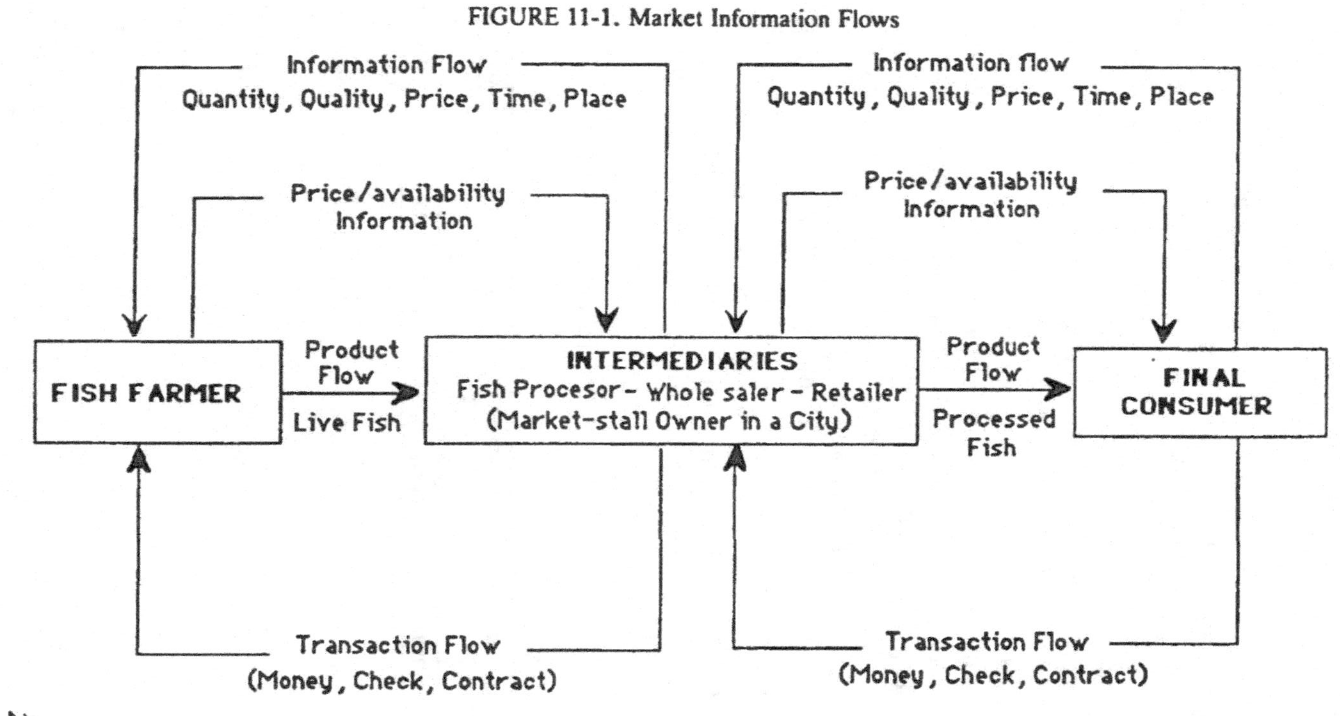

transaction does not necessitate an intermediary, or the farmers might develop their own capability for handling fish. The complexity of alternative marketing channels is illustrated in Figure 11-2. In A, there is a direct sale. In B, the fish farmer goes through a retailer, whereas in C an additional stage occurs. The most complex channel is D. One might wonder whether the transfer of the product through the numerous stages does not lead to inefficiencies and higher cost to fish farmers and consumers. However, if the intermediaries handle fish carefully and efficiently, and provide the desired quantity and quality of fish product on a timely basis, consumers are satisfied with prices and the farmer may also realize increased earnings. It must be remembered that the intermediary transfers not only a product, but also needed information in product quality, quantity, availability, and price. If intermediaries handle fish improperly and sell a poor quality product at high prices, they may undo all marketing efforts of the fish farmer. Choosing the right marketing channel is essential to successful marketing.

As the fish move through the channel, the price at each stage is increased in accordance with value added to the product. Fish

FIGURE 11-2. Distribution Channels for Fish

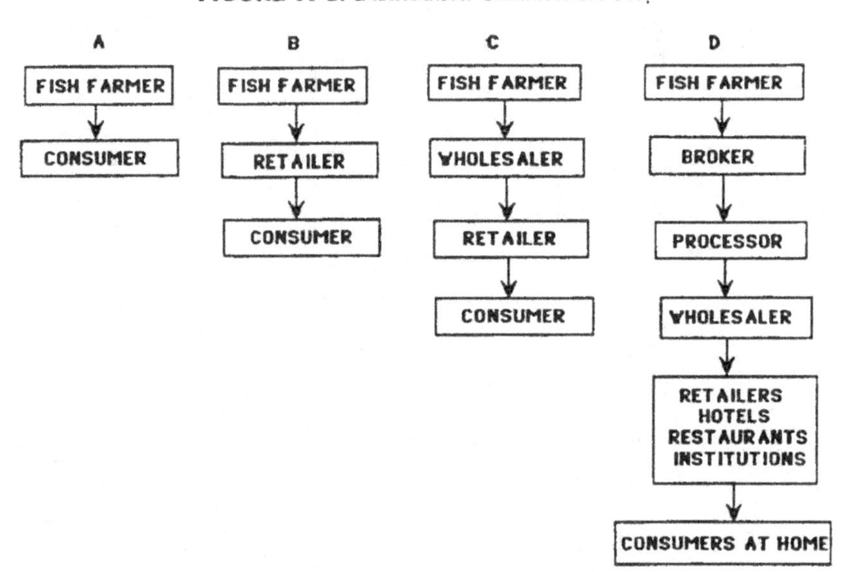

farmers are not always pleased with the difference in the price they receive and consumers pay for the product. The difference, called the *marketing margin*, reflects the percentage level of services provided at each stage through which the product passes, and the levels of demand and supply at different times of the year. An examination of typical prices of catfish received by farmers and processors in the U.S., depicted in Table 11-1, gives an example of marketing margins. In January, there was a $1.39 difference from producer price to consumer price. This represents a 78 percent increase in prices for service rendered. In June, the difference was $1.38, representing an 84 percent change. The December change was $1.37, representing a 100 percent margin. Obviously, the marketing margin is affected by time of sale and price paid for the raw product. The proportion of price received by processors increased even though the price received fell.

Many other factors also have direct impacts on the margin. For example, governmental price controls, producer organizations, type of product, and market concentration, plus many other factors may alter the margin at any given time or place.

Length of Channels

Fish distribution channels may be quite simple or very long and complex. The length and complexity depend on the volume of fish moved, the number of functions performed, scale of operation at each stage, and the distribution system chosen. The organization of the distribution will depend on the type of markets, and the organization of fish farmers.

TABLE 11-1. Typical Prices Received by Farmers and Processors for Catfish, Alabama, 1989

Business type	January	June	December
		dollars	
Price received by farmers	78.0	75.0	68.0
Processor price	217.3	212.9	204.6
Processing and marketing margin	139.3	137.9	136.6

Source: The data were taken from *Aquaculture and Situation Outlook Report*, USDA, September 1989.

Increasing Market Share

Commercially produced fish are in direct competition with those produced in the wild and from imports. In order to effectively compete in the fish market, the market manager of commercially produced fish must isolate the variables that affect sales, as well as those that may be affected by the action of the producers. The market manager may increase market share by monitoring the following variables:

1. product;
2. price;
3. promotion; and
4. place.

Product

The product market manager knows that the key in expanding sales is a premium product. Therefore, the manager will have products examined and inspected carefully before sale. The manager also will ensure that size, taste, packaging, and other characteristics pleasing to the consumer are satisfied before putting the product on the market. Off-flavor, color, texture, and general appearance are key elements that must be monitored carefully since product quality improvement is a good means of increasing market share of any product.

Price

To minimize price, the manager will keep processing costs at a minimum to ensure that product price is competitive. Price in a purely competitive market situation is determined by the forces of demand and supply. The market is cleared when market equilibrium is reached. This is particularly the case in a purely or nearly pure competitive market situation. In a monopoly, the seller determines the price at which each quantity is sold. This process is called *price determination*. In product marketing, price is also set by a system called *price discovery*. This is a process by which buyers and sellers arrive at a specific price for a given lot of produce in a given location. The demand and supply price target must be discovered and applied to each transaction in the market place. In the aquaculture industry, producers and processors bargain on a price which is usually close to the prevailing market price. Because of the highly competitive nature of the industry, both parties know that an error in pricing may have serious consequences for the participants. Therefore, the fear by sellers that processors are acting "unfairly" often leads to other forms of price determination, such as cooperatives or marketing associations, in which producers "collude" on price and market conditions.

Promotion

Promotion involves programs to encourage sale and increase market share at any point in the channel by influencing potential purchasers. The role of promotion is to communicate information including product quality, price, and benefits of the product to potential clients. Promotion acts on both the intellectual and emotional state of the buyer. Competition is keen throughout the channel, and promotional effects may be necessary just to maintain market position for a species or for a specific group of producers. In developed market channels, "food fairs" and other sponsored events are held to display both new and traditional fish products to clients for evaluation prior to full-scale promotional campaigns.

Place

Distribution of the product to locations used by customers wishing to purchase the product is a market function. In the United States, catfish farmers transport fish to processing plants up to 50 miles from the ponds. The processed fish may be transported an additional 500 miles before final purchase takes place. Frozen products are transported the greatest distances. In many developing nations, dried fish are transported great distances because of the lack of refrigeration both in transport and local storage. Producers and processors may increase market shares by transporting fish to locations previously inaccessible to consumers.

INTEGRATION

Integration in marketing includes all the ties which relate the ultimate consumer to the resources that are to be employed in producing the services or goods demanded. It embraces the various means of achieving coordination for the optimal operation of two activities.

There are many types of integration in the agricultural sector. The two types most frequently encountered are vertical and horizontal. Vertical integration is defined as that type of organization that comes into existence when two or more successive stages of production and/or distribution of a product are combined under the same control. A vertically-integrated firm in the fish industry is one in which the owning, managing, or controlling company may direct all or portions of the hatching operations, grow-out operation, manufacture of feeds, processing capability, and the marketing of the product. In Table 11-2, stages of vertically-integrated system aquaculture are shown.

A newly-developing industry frequently demonstrates a high degree of vertical integration. As the scale of operations grow for both the industry and firms, the firms give up the functions which may be performed more efficiently by independent firms. In the early development of the U.S. catfish industry, fish farmers produced their own fingerlings, fish, and feed, and then marketed their products. As industry growth occurred and the scale of operations expanded, farmers and participants in the industry began specializing, although many farmers and firms are still highly integrated. The firm is said to be fully integrated when its activities are extended to include some equipment or some type of organizational setup at all production and distribution stages that are normally included in the industry in which the firm is operating.

Horizontal integration occurs when two or more similar concerns are combined to perform the same functions in the same stage of distribution or production. It is the process through which firms grow laterally by gaining control over other firms performing similar activities at the same level in the marketing sequence. One form of horizontal integration is the consolidation of several small producers under a single management.

TABLE 11-2. Stages of a Vertically-Integrated Marketing System in Aquaculture

Stage			
SUPPLIES	Feed ingredients chemicals floating feed sinking feed antibiotics	Consumables and gadgets oxygen meters PH testing equipment	Equipment and farm buildings tractors feeders holding tanks
FARMING		brood stock,egg and fingerling production hatchery nursery growing	
PROCESSING		handling dressing packing freezing smoking canning salting storing	
DISTRIBUTION		brokerage promotion wholesaling retailing	

Marketing analysts argue that market-related horizontal integration has been a more successful route to growth than vertical integration. By extending the range of goods and services offered to a single marketplace, concerns may exploit their established commercial skills. This may be especially true in aquaculture. The skills necessary to produce fish are far different from those in processing or distribution. Successfully combining all the processes vertically requires exceptional management skill. On the other hand, horizontal duplications of effort and pond or processing facilities are much easier to accomplish.

Integration of firms within the industry results in economies for the firms involved. There are three kinds of situations under which firms will operate to lower costs:

1. *Functional Efficiency*. The firms might be concerned about optimum scale for the performance of a series of vertically-related functions such as:

a. supply of material and service;
b. farming, breeding, and growing fish;
c. processing; and
d. distribution and final sale of the fish product.

2. *Optimum Managerial and Technical Units*. If the optimum managerial unit is larger than the optimum technical unit for production, integration will be encouraged. Suppose that the most efficient size for a fish feed manufacturing company is not great enough to absorb the talents of a trained and experienced manager; management could be used to full capacity by extending operations over fish production units.
3. *Coordination*. A major reason for integration is that of coordination. An integrated aquacultural firm affects coordination by administrative action, while the non-integrated firms are coordinated through the market. Through coordination, information is transmitted more readily and the degree of uncertainty in the operation of the vertically-related enterprises is decreased.

CATFISH MARKETING CHANNELS IN THE SOUTHERN U.S.

Catfish marketing developed initially on a localized basis with supply being dependent on wild fish captured from rivers and lakes. In 1969, wild catfish comprised approximately two-thirds of the total amount sold. Today most catfish marketed come from high density culture systems such as cages, tanks, raceways, and farm ponds. Some limited sales within the United States are from wild catches. The rest come from imports primarily from Brazil, most of which is wild catch. The secondary import source of ocean catfish represents several species and is an entirely separate product.

In the early days of catfish culture, fish and fish product marketing was done directly to consumers at the farmgate. The farmer used roadside tanks in which fish were placed at a convenient location contiguous to the farm. A sign was simply placed to indicate "catfish for sale." Passersby stopped their vehicles and purchased fish live or processed. It was estimated that only a limited quantity (approximately 1800 pounds per farmer) was sold by catfish farmers through this channel. Usually a processing fee of 25 to 30 cents per pound was charged for processing services.

Another channel which has experienced rapid expansion is that of "fish-out." Fish-out is a recreational activity in which individuals generally pay a fee for fishing and also pay a price for the fish caught. Producers range from farmers with a sign and an honor box for fee deposits to operators who provide camping, picnic tables, and concession stands. Pond location relative to population centers is a primary factor in fish-out operation profitability. The quantity sold through this operation is vastly increasing, but still very limited. The amount of fish moved per farmer through this channel is estimated at 16,000 pounds annually. A substantial portion of the fish sold by commercial producers to live haulers is ultimately marketed through fish-out. Live haulers transport fish for stocking owned ponds or for sale as stocks to other operators.

The bulk of commercially-grown catfish today is sold to processors. The traditional method has been to drain some water from the pond and seine-harvest fish from the catch basin for sale to processors. Today, however, many ponds are drained infrequently. They are constructed in a diked fashion which permits seining without lowering water levels. This process also reduces much of the pond waste effluents dispersed into nearby waterways.

Processing plants range in size and capacity and may be owned by single individuals, companies, or cooperatives. The processors then sell to brokers who market fish to wholesalers, retailers, and institutions. Eventually the fish are sold to consumers. The fish may undergo several form changes throughout the channel. They may be processed and transformed into several products such as fish cakes, fillets, nuggets, etc. The marketing network for the catfish in the United States can be exemplified by Figure 11-3.

FIGURE 11-3. Schematic Representation of Catfish Distribution Channel for Southeast U.S.

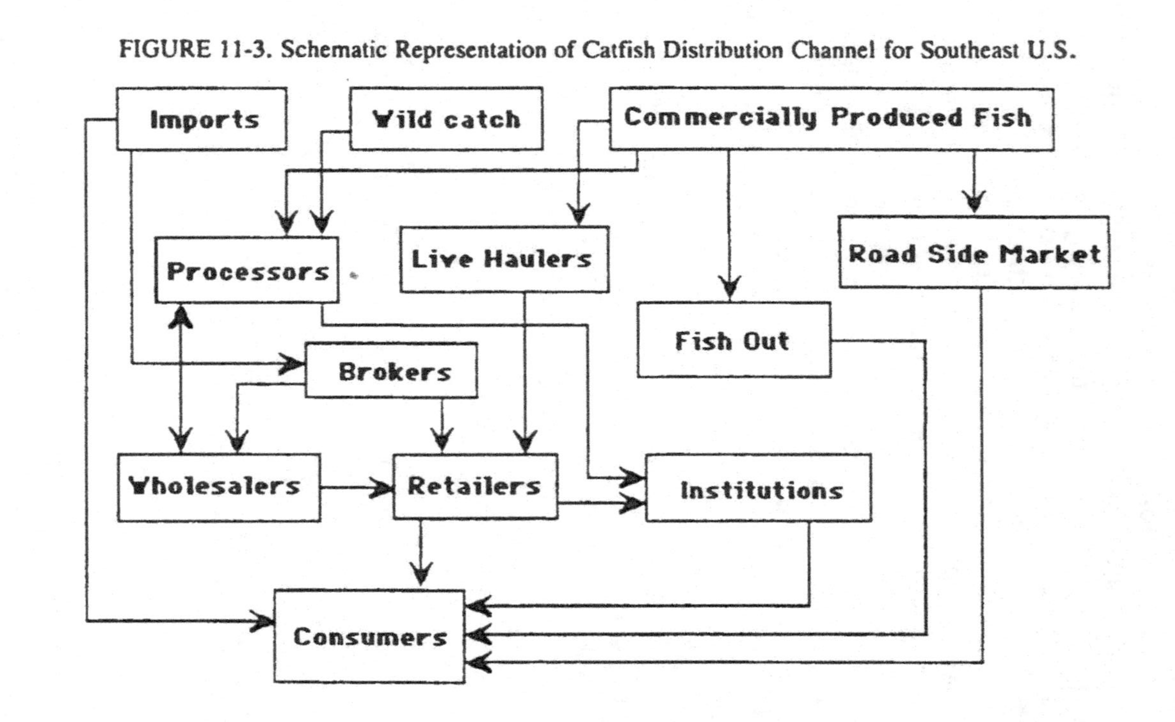

REFERENCES AND RECOMMENDED READINGS

Abella, C. 1986. *Marketability of Tilapia in Metro Manila; in Tilapia Farming*. Proceedings of the First National Symposium and Workshop on Tilapia Farming. PCARRD, Los Baños Laguna 40-2.

Adams, C.M., F.J. Prochaska, and T.H. Spreen. 1987. "Price Determination in the U.S. Shrimp Market." *Southern Journal of Agricultural Economics*. 19, 2:103-11.

Anderson, J.L. 1985. "Market Interactions Between Aquaculture and the Common-Property Commercial Fishery." *Marine Resource Economics*. 2(1):1-24.

Cacho, O., H. Kinnucan, and S. Sindelar. 1986. *Catfish Farming Risks in Alabama*, Circular 287; Alabama Agricultural Experiment Station, Auburn University, Auburn, Alabama. 19 pages.

Campbell, M. 1974. *Stable Tropical Fish Products*. Report on a workshop held in Bangkok, Thailand. 27 pages.

Chaston, I. 1984. *Business Management in Fisheries and Aquaculture*. England: Fishing News Books, Ltd. 128 pages.

Chaston, I. 1983. *Marketing in Fisheries and Aquaculture*. England: Fishing News Books, Ltd. 143 pages.

Engle, C.R., U. Hatch, S. Swinton, and T. Thorpe. 1989. *Marketing Alternatives for East Alabama Catfish Producers*. Alabama Agricultural Experiment Station, Auburn University, Auburn, Alabama, Bulletin 596. 27 pages.

Food and Agriculture Organization of the United Nations. 1987. *Women in Aquaculture*. Proceedings of the ADCP/NOKAD Workshop on Women in Aquaculture, Rome, FAO, pp. 13-16.

Glude, J.B. 1983. "Marketing and Economics in Relation to U.S. Bivalve Aquaculture." *Journal of World Mariculture Society*. 14:576-86.

Guevera, G. 1987. *Handling and Processing of Tilapia; in Tilapia Farming*. Proceedings of the First National Symposium and Workshop on Tilapia Farming. PCARRD, Los Baños Laguna, pp. 43-7.

Hempel, E. and S.S. Van Eys. 1985. "The Shrimp Industry – Markets, Quality Problems and Competition from Aquaculture." *Infofish Marketing Digest*, pp. 22-6.

Hottlet, P. 1982. "Selling Shrimp to the EEC." *Infofish Marketing Digest*. May, 15-19.

Kinnucan, H. and D. Wineholt. 1989. *Econometric Analysis of Demand and Price-Markup Functions for Catfish at the Processor Level*. Alabama Agricultural Experiment Station, Auburn University, Auburn, Alabama, Bulletin No. 597, page 30.

Liao, D.S. 1984. "Market Analysis for Crawfish Aquaculture in South Carolina." *Journal of World Mariculture Society*. 15:106-7.

Lin, B. and M. Herman. 1988. *An Economic Analysis of Atlantic Salmon Markets*. Proceedings of the Symposium for Seafood and Aquacultural Products; International Institute of Fisheries Economics and Trade and the South Carolina Wildlife and Marine Resources Department, pp. 72-84.

McCoy, H. 1988. "Aquaculture and the Stock Market." *Aquaculture Magazine*. January/February, pp. 20-30.
Miller, J.S., J.R. Conner and J.E. Waldrop. 1981. *Survey of Commercial Catfish Producers: Current Production and Marketing Practices and Economic Implications*. AEC Research Report No. 129, Agricultural Marketing Service, USDA Mississippi Dept. of Agriculture and Commerce, page 20.
Mims, S.D. and G. Sullivan. 1984. *Improving Market Coordination in the Catfish Industry in West Alabama*. Bulletin No. 562, Alabama Agricultural Experiment Station, Auburn University, Auburn, Alabama, page 31.
Pigott, G.M. 1988. *Radurization of Aquaculture Fish: A Value-added Processing Technology of the Future*. Proceedings of Aquaculture International Congress and Exposition; Vancouver Trade and Convention Center; Vancouver, British Columbia, Canada, September 6-9; pp. 181-7.
Rameses, Jr. S. 1987. *Tilapia Marketing in Metro Manila; in Tilapia Farming*. Proceedings of the First National Symposium and Workshop on Tilapia Farming PCARRD, Los Baños Laguna, pp. 34-9.
Shaw, S.A. 1986. *Marketing the Products of Aquaculture FAO Fisheries*. Technical Paper 276, Food and Agriculture Organization of the United Nations, Rome, page 106.
Shepherd, G.S. and G.A. Futrell. 1982. *Marketing Farm Products*. Ames: The Iowa State University Press, pp. 428.
Smith, I.R. and K. Chang. 1984. "Market Constraints Inhibit Milkfish Expansion in Southeast Asia." *Aquaculture Magazine*. September/October, 1984, pp. 28-33.
USDA, 1989. *Aquaculture Situation and Outlook Report*. Economic Research Service, AQUA#3, p. 34.
Vondruska, J. 1984. "U.S. Consumer Attitudes Toward Fish." *Infofish Marketing Digest*. 3:30-34.
Whetham, E.H. 1972. *Agricultural Marketing in Africa*, London: Oxford University Press, page 240.

Chapter 12

Government in Aquaculture

Aquacultural activities throughout the world have long been influenced by governmental intervention. Intervention has had many forms including: direct price subsidies to producers, consumer price ceilings, pond construction subsidies, water supply provisions, transportation discounts, tax policies, and, negatively . . . subsidies to marine fish industries. More recently, another form of intervention has been the tendency of government to either totally ignore or strictly enforce regulations related to environmental quality, particularly water quality concerns.

In the United States where the aquaculture industry is relatively young, there is little history of direct government intervention. Indirect effects, however, are seen in many aspects of the aquacultural industry. For example, the impact of the grain policy is seen in fish feed prices. There are regulations on antibiotics in feed. Income tax rulings affect the depreciation of farm ponds and equipment. Additionally, there are health and safety standards in the processing sector. Yet, the overall influences of governmental activity regarding aquacultural producers in general seem to be unnoticed. A point may be reached, however, where management decisions must be made in light of what may become governmental policy in the foreseeable future. To understand the implications of alternative political policies which may arise, one must understand the framework within which such policies must operate. Generally, policies are reactive steps of government to problems within a particular economic sector, such as agriculture, or in this case, aquaculture.

TYPICAL AQUACULTURAL PROBLEMS

Aquacultural products are primarily food sources. Consequently, much of the governmental activity worldwide has focused on improving the nutritional well-being of the indigent population by increasing food (agricultural) output through aquacultural products. In some developing countries, fishery products may contribute over half of the total animal protein annually. Because some of those products also have high value for foreign exchange, there has been an income effect associated with growing aquatic animal species. Thus, aquaculture has been quite popular as both a food and an income source for much of the developing world. For example, Sierra Leone in West Africa is heavily dependent on fish products as a protein source, but oyster culture is now beginning to emerge as a primary income source (Kamara, 1982). Research in Malaysia has shown the feasibility of polyculture systems for combining food (tilapia and carp) and cashcrops (prawns) in the same freshwater ponds (Clonts et al., 1989).

The decrease in yields from marine and inland capture fisheries worldwide has increased the attention on aquaculture. In Nepal, for example, about two-thirds of the total food fish product is contributed by aquaculture; in both China and India, over 40 percent is derived from fish farming (Shang, 1990). In the Philippines and Indonesia, serious marine fish population problems attributable to overharvesting have been documented. These nations are increasingly turning to aquaculture as alternative food sources. Obviously, the degree of access to the seas for marine fishery products is a factor in the intensity of inland fish culture. Yet, as depletion of the wild fishery continues, attention will be turned increasingly to culturing aquatic species to maintain protein and caloric levels for the general population. With increased attention to maintaining food and income sources, one should expect greater impacts of one particular agricultural/aquacultural problem worldwide . . . supply control.

Supply Control

World aquacultural producers typically find themselves involved with the catch-all problem labeled "supply control." Typically, producers do not react well to price level changes except to stop production when price falls below operating expenses for too long a period of time. Much of the world aquaculture production is conducted on a subsistence level in rural atomistic societies such as may be found in impoverished areas of developing countries. Mexico, China, and other Asian nations provide numerous examples of this condition (McGoodwin, 1982). Alternatively, cultured products also may be viewed as a supplement for producer income. Typically in those cases, unpaid family members provide the labor and management to produce that extra financial boost to the family or community. Hence, low profits associated with low prices tend to have only limited impact on supply. Additionally, in many cases where subsistence or supplemental production exists, high prices do not necessarily entice additional supply, nor do low prices drive supply downward. This problem of "irreversible supply" occurs because there is neither room to expand production, nor incentive to sell the product rather than consume it. Obviously, exotic species such as shrimp (prawn), oysters, and others cannot be classified as low profit production items, but the general condition prevails for producers of most aquacultural products.

In more developed economies such as Europe and North America, aquaculture enjoys a status position in that producers are more successful in achieving a quality life through fish production. In fact, aquaculture was the enterprise that many United States farmers, especially those in the Southeast, turned to when traditional agriculture failed in both price and production during the 1980s. Yet, even in the more developed economies, the prevailing problem for producers is their inability to deal with rapidly changing prices for both inputs and products. Asset (capital and land) fixity, credit dependent production, and non-transferable management skills tend to keep production going even when rational economic thought would dictate temporary resource reallocation.

CAUSES FOR AQUACULTURAL PROBLEMS

Water Supply and Quality

Aquaculture naturally is water-dependent. Hence, the industry is affected by any event that affects the water supply. Drought, pollution, floods, etc., must be dealt with on a regular basis. Everybody talks about the weather, but who can control it? Weather and water supply are not the chief concern of aquaculturalists. Rather, they are just accepted as variables in the overall production scheme. Other economic problems are more pressing.

Among these more pressing issues is the variable of water quality that in the past has been generally ignored, but must be faced in the immediate future. To be sure, the quantity and quality of the water supplied to aquacultural operations is a key factor in production. The output side of the operation, however, is also critical. Naturally, the fish farmer prefers to think in terms of the liability damage others are imposing on fish production. Yet, that same farmer must realize that those fish are growing in what some may consider a "sewage lagoon." The fish pond contains all the waste feeds, animal excrements, plant materials, and even pharmaceuticals or other chemicals used in fish production. When pond waters are released into receiving streams which are later used for domestic consumption or other purposes, the impact may be dramatic (Hebicha, 1989). Thus, there are two sides to the water quality issue in aquaculture. Ives (1989) reports that pollution from both shell and finfish aquaculture can be effectively controlled and even turned into an advantage with proper farm management techniques. Yet, resistance by aquaculturalists to changes in production technology may result in strong negative public opinion against the operation. Producers need to be evaluating every aspect of the pollution problem so as to prevent negative reactions and turn the issues into marketing gains. For example, good ideas include introducing traditional agricultural practices into aquaculture such as crop "rotation" (i.e., moving fish cages around), waste management lagoons, or using shellfish filter feeders as indicator species of environmental conditions.

Recent large oil spills in Alaska, Texas, Italy, and the Persian Gulf quickened the realization that the existing and potential coastal mariculture industry is in constant threat of disaster. What policy, what government program, is present to protect those interests? Actually, there are some, but economic redress may not come without legal action. Hence, the aquaculturalist may pay more in legal fees than could be gained in punitive and liability damages. Such is the state of aquacultural production in much of the world, whether it be salt water mariculture or fresh water aquaculture.

Competition for high quality water is a problem facing world aquacultural expansion. This is particularly true for hatcheries. Aquaculture is likely to compete with traditional agriculture, municipal needs, and other industries for water supply. At present, the world is approaching a crisis with respect to the potable water supply. If the trend continues, there are likely to be user conflicts for that supply.

Farmers in much of Asia are concerned about the quality of water available for fish growth. In coastal regions, excessive withdrawals for aquaculture are causing salt water intrusion and subsequent pollution of aquifers. Inland, multiple uses of water for irrigation and housing as well as for aquaculture contribute to deterioration of ground and surface water supplies. Typical aquaculture production schemes include rice cum ducks, rice cum fish, or even rice cum housing, and several other combinations. Each of these systems has provision for some minimal and perhaps even large doses of chemicals, or other toxic materials to be introduced into the aquatic ecosystem. Additionally, there is ample evidence of self-pollution. Accumulation of feed and animal wastes, chemicals, etc., in reservoirs and pond outfalls can create localized pollution problems of significant magnitude. A cursory examination of governmental programs in much of the Third World shows limited concern with problems of this nature. Yet, the population is dependent on the fish as a general food and protein source.

Price and Income Conditions

At issue throughout the world is farmer income. Elasticity, discussed in Chapter 3, is the measure of a percentage change in quantity demanded associated with a 1 percent change in the price of a good (own or price elasticity) or consumer income (income elasticity). Usually, aquacultural products, namely fish, have farm-level price elasticities of less than $|1.0|$. In other words, they are price inelastic, meaning that sales of the product do not respond greatly to changes in price. Kinnucan et al. (1988) estimated that the farm level price elasticity for catfish in the Southeastern United States was around 0.37. This means that if prices rise 1 percent, farm sales will fall only 0.37 percent. On the other hand, it also means that if prices fall, consumers will not rush out to stock up on the products because at the processor level demand was elastic −1.54 in the Alabama study. At the retail level it reached −2.5, reflecting the degree of substitutability of other foods for catfish. At issue is just how much fish the typical family desires to have in the daily diet.

Fish consumption with respect to income varies significantly throughout the world. For example, in the United States, fish is income elastic, meaning that as incomes rise, more fish is consumed. In Malaysia, fish is income inelastic, especially for the species most easily produced (Alsagoff et al., 1990). Malay consumers substitute poultry and beef for fish as incomes rise. In each case, however, the long-run consumer patterns are of utmost importance to fish producers. Yet, it is the short-run decisions that wreak havoc in the markets and force producers into financial difficulties. Hence, there may be justification for government intervention to stabilize markets, create new markets for existing marine products, and greater markets for various cultured species. Taken together, price and income conditions have much to do with aquaculture. The framework of management decisions can be and is strongly influenced by governmental policy action.

Biotechnology

The aquacultural world is on the brink of major change brought about by test-tube experiments. Genetic studies and hybridization have introduced new products which exhibited growth and vigor not seen in wild animals. For example, the Blue x Channel catfish cross produced a much more highly desirable fish than either of the parents. Triploid varieties of the white amur are now being used for aquatic weed control in large impoundments, although numerous concerns are being registered over the ultimate effects. Now, genetic engineering is taking place. Gene splicing is showing promise as an exotic means for controlling growth and disease as well as reproduction. Transgenic fish or fish altered via use of human growth hormones are raising fears of product safety as well as biological security. A number of antibiotics are being tried for the control of diseases in shrimp production (Williams and Lightner, 1988). What are the safe limits? How much increase in the use of antibiotics, hormones, and other chemical substances is practical before residues begin to affect human health? What long-term unseen impacts await us if these practices are implemented worldwide? What role is government to play in this drama? What legislation is needed to protect both the human consumer and the world environment?

Asset Fixity

Fish producers face one chronic problem seen everywhere in agriculture. What does one do with a used or undesired fish pond, or obsolete fish processing equipment? The question is very much like that of what to do with a large combine when soybean prices are at an all-time low. That resource is either used or left exposed to the elements of weather. Generally, the choice is to use the asset and hope for better conditions next year. If prices are low, is it a sign of overproduction and too many fish ponds? Perhaps so, but to the individual producer, the choice is obvious. Because of sunken cost and long-term debts, the only recourse often is to continue production. This means that prices may go even lower than average unit total cost and bring about worse economic conditions. Why do farmers and aquaculturalists continue to produce? First, the capital used in production has an opportunity cost (alternative productive value) near zero. Second, fish production has become a way of life or a lifetime investment, and third, lack of skills for other forms of employment for the producer and his family frequently give them no choice but to continue production. Is there hope for those in this situation through government intervention?

Equity in Income Distribution

A real danger of government intervention in a particular market is that the benefits will accrue to individuals who bear the least cost related to the problem a program is designed to resolve. Inequities are difficult to foresee, but political arrangements and influences do provide the environment in which they may occur. An easy example is found in the United States agricultural commodities program over the years. Quotas, price supports, and crop restrictions, etc., were designed to apply universally to all eligible farmers. However, the more successful farmers were able to capture the larger benefits and at the same time minimize production risks. On the other hand, less productive farm units were able to utilize program benefits primarily to stay in business when the opportunity cost for doing so was quite high.

Generally, governmental intervention is a political activity. Thus, there may be several agendas hidden in the program. For example, the agricultural/aquacultural credit programs of Malaysia do provide assistance to farmers, but they also are used to hold consumer prices down (Alsagoff et al., 1990). Throughout Latin America, problems of the burgeoning urban populations may be approached via various programs directed toward the rural sector. However, the urban poor tend to gain at the expense of the rural poor unless the programs clearly apply equity principles in planning and implementation. Actually, this is a chronic problem worldwide. The magnitude of problems, especially in developing nations, forces governments to combine several goals under one project, with the result being a program that tends to foster inequities in areas where the program "beneficiaries" are least able to afford the costs.

Programs to Resolve the Issues

As indicated in the beginning of this section, governments worldwide have attempted to "handle" the issues of price and production under a variety of schemes. These schemes typically fall into six categories which directly affect producers, and into at least three others which have indirect effects:

Direct Impacts

- Price supports
- Production limits
- Support payments
- Producer credit
- Research and extension programs
- Demand expansion

Indirect Impacts

- Consumer price controls
- Environmental quality
- Public works projects

Direct Impacts

Direct impacts such as price support policies and programs have a long history. They are the most elemental form of governmental intervention in a free market economy. Basically, they are subsidies paid to fish farmers either directly as cash bonuses or indirectly as setting minimum market price for the good. There has been limited application of price supports for aquacultural products in the United States, but there are instances of support in other nations such as Indonesia, Malaysia, and Israel.

Production limits often coincide with price supports because quotas or other production restrictions are imposed to decrease supply and force prices to a higher level. In developing nations, the concept of restricting production may have limited appeal for several reasons, foremost of which may be concern over consumer prices and the need for ample protein sources. Consequently, production limits typically would not be introduced unless there is already a problem with excess supply, not concern with price levels.

Support payments typically do not occur unless market price falls below some stated "target" price established by the government. In other words, the government will make up the difference between the prevailing market price and the "target price" if necessary. Alternatively, support payments may be in the form of supplements to produce a particular crop or fish species, undertake certain conservation practices, or perhaps continue production of new, staple, or different products for which the current market is too weak or small to support the production unit (farm).

Credit policies frequently are a means to give support without making it appear that large sums of money are being invested in a particular segment of the economy. Loans to subsistence farmers for new pond construction or new water source projects are means to reduce the risk of the individual entrepreneur and provide additional production in the short-run when economic conditions would dictate otherwise. Governmental loan programs with favorable credit terms are a popular means to offset the negative effects on producers of market price ceilings imposed to insure relatively low consumer prices. Examples of this sort of intervention are seen throughout Asia and Africa in countries such as Malaysia, Nigeria, and others.

Research and extension programs provide new technologies for existing producers and for attracting new ones. Research is expensive. Consequently, most research is performed by the more developed nations of the world (although perhaps done in a host developing nation). Results of research are then disseminated to producers in developing areas through local extension programs.

Demand expansion is always a desired goal of government planners. If the demand for fish products can be increased domestically, and especially internationally, many problems facing producers will appear to fade; for example: producer incomes and levels of living may rise, new technology to increase production may be adopted, protein intake of the population may increase, and so on one may go in claiming benefits. Yet, in many instances, demand expansion efforts are merely an alternative way of introducing highly competitive practices into a specific market. On the other hand, long-term, deliberate programs to change tastes and preferences of consumers toward products in abundance may be more effective. In Malaysia, the opportunity is present for significant expansion of tilapia production (Clonts et al., 1989). However, tilapia are not highly considered by the population. Farmers prefer to produce prawns for export. Simply increasing prawn production to stimulate greater foreign exchange is not going to resolve the need for protein sources among the rural poor. But, introducing different tilapia species such as those of different color and appearance may change preferences enough to expand the demand and give producers higher incomes as well as better nutrition.

Indirect Impacts

Indirect effects may be more likely realized from policies to protect or enhance consumer well-being than those reflecting producer needs. Price ceilings, environmental quality controls, and public works projects may all have either positive or negative impacts on aquacultural production. But, there is reason to suspect that more negative than positive impacts may be realized by producers. Consumer protection policies may be found almost anywhere. Israel,

Costa Rica, Panama, and The Cameroon are examples of the many nations that have enacted price ceilings in aquaculture markets.

ECONOMICS OF SUPPORT POLICIES

Price supports, loan programs, direct support payments, and even credit allowances all amount to some sort of guarantee for the producer in the short-run. Since risk is a major element in limiting production, programs to reduce risk will result in increased production, unless tempered by limits on total production, for example, acreage (hectare) restrictions or limits on tons of fish produced.

History has shown that price supports likewise tend to cause supply to exceed demand because the typical support price or target price exceeds that necessary to clear the normal market. Consider Figure 12-1. If the price to clear the market under normal production conditions is P_m and the market quantity is Q_m, any higher price will result in greater market quantity being supplied, $0Q_2$. The higher support price likewise results in less quantity demanded, $0Q_1$, so that a market surplus, Q_1Q_2, occurs ($Q_2 > Q_1$). This is not necessarily bad, but it indicates that there is a cost, in this case to consumers, for any form of price control.

Costs may be shifted to taxpayers in general through alternate policies using the same model. If the support price is a "target price" guaranteed after market clearance, the same net effect with respect to the original market quantity is possible. In this event, price P_1 is paid to producers as a "bonus" above the market price P_m, and there is no surplus since the normal market quantity desired, Q_m, is purchased by consumers. The higher income for the producer is reflected in the shaded area of Figure 12-1. The private gain by producers represents additional social costs which must be made up in some other area of government policy, possibly through taxes on consumer incomes or on total purchases (sales tax).

Production controls, on the other hand, limit supply in order to drive price upward (see Figure 12-2). Production limitations such as production quotas, acreage (hectare) limits, or other types of land reserve programs are designed to limit supply. The original market price and quantity, P_1 Q_1, theoretically is shifted upward and leftward along the demand curve to a new equilibrium, P_2 Q_2. How-

FIGURE 12-1. Economic Effects of Price Support Programs

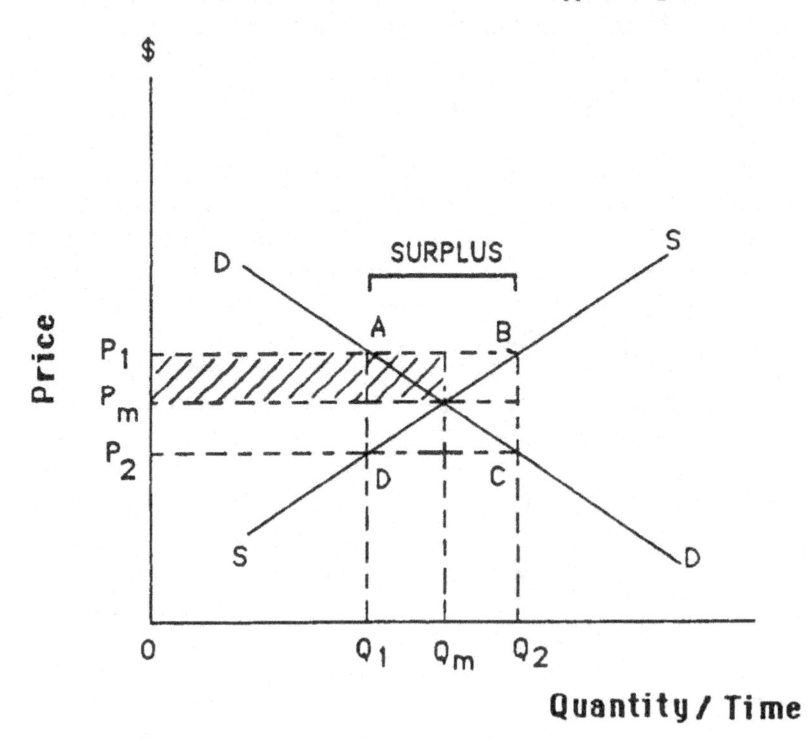

ever, producers should be expected to use existing and new technology to increase production and actually increase supply, especially if there is a price support program simultaneously in effect. The result could easily be supply S_3 (Figure 12-2) with subsequent lower prices and increased quantity. If the control mechanism is a quota or land use (crop or acreage) allotment, there is a tendency for the extra income gain derived from any support price to be incorporated into the land, or into the actual allotment if a market for transferring quotas or allotments is created. Consequently, those holding the original quota may receive a windfall gain in the form of higher income, higher land value, or a monetary gain via market value of the quota itself. That was the result of United States policy on commodities such as cotton, tobacco, and feed grains between 1950 and 1980. Production limitations without punitive sanctions on pro-

FIGURE 12-2. Economic Effects of Supply Control Programs

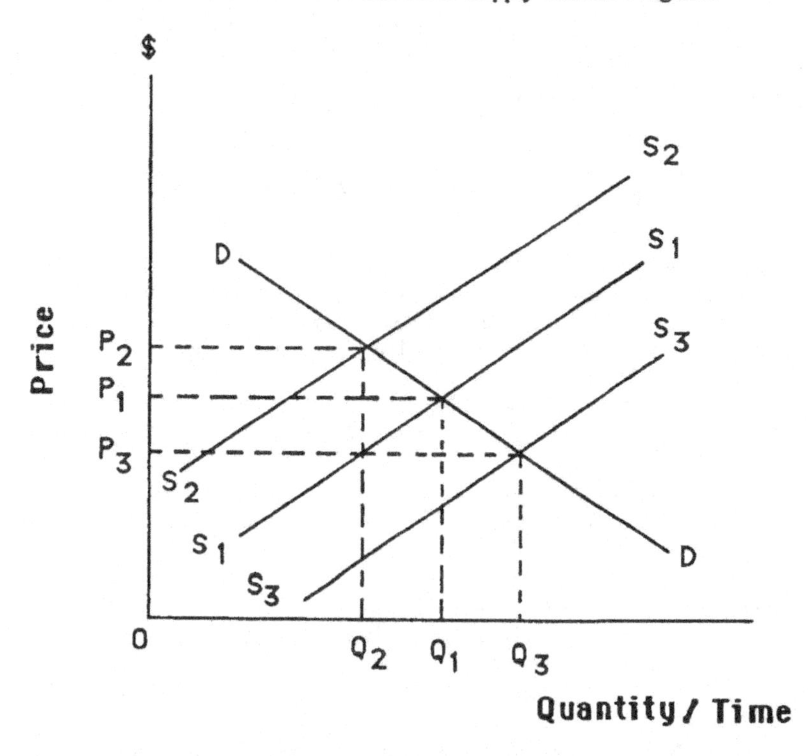

ducers generally will not be effective. There is too great an incentive for individual producers to violate the restrictions in order to take advantage of any general price increases.

The conclusion of this discussion is that governmental intervention into the market for agricultural or aquacultural commodities is actually market manipulation. Either price or quantity is manipulated or controlled in some way to achieve the goals of government for either producers or consumers. The specific form of controls used in policy should be carefully evaluated prior to implementation since if improperly used, the policy could result in opposite effects than those intended, as shown in the preceding two examples. Also, most policies of this nature should be viewed as short-term in nature. One mistake frequently made is for the government to continuously re-authorize the same type of short-term solutions for long-term problems. Instead, there should be longer-term solutions planned and short-term policies enacted to guide production and consumption along the lines needed for ultimate social and economic good of the economy.

REFERENCES AND RECOMMENDED READINGS

Alsagoff, S.A.K., H.A. Clonts, and C.M. Jolly. 1990. "An Integrated Poultry, Multispecies Aquaculture for Malaysian Rice Farmers: A Mixed Integer Programming Approach." *Agricultural Systems*. 32(1990):207-31.

Brown, J.H. 1989. "Antibiotics: Their Use and Abuse in Aquaculture." *World Aquaculture*. 20,2:34.

Charmantier, G., M. Charmantier-Daures, and D.E. Aiken. 1989. "Accelerating Lobster Growth with Human Growth Hormone." *World Aquaculture*. 20,2:52.

Clonts, H.A., C.M. Jolly, and S.A.K. Alsagoff. 1989. "An Ecological Food-niche Concept as a Proxy for Fish-Pond Stocking Rates in Integrated Aquacultural Farming for Malaysia." *Journal of the World Aquaculture Society*. 20,4:268.

Hebicha, H. 1989. "Catfish Pond Harvesting Pollution Control and Costs for West Central Alabama." Unpublished Ph.D. Dissertation, Department of Agricultural Economics and Rural Sociology, Auburn University, Alabama.

Ives, B.H. 1989. "Pollution! What's a Farmer to Do?" *World Aquaculture*. 20,3:48.

Kamara, A. B. 1982. "Oyster Culture in Sierra Leone," in *Aquaculture Development in Less Developed Countries*, L.J. Smith and S. Peterson, eds. Boulder: Westview Press, pp. 91-108.

Kinnucan, H., S. Sindelar, D. Winehold, and U. Hatch. 1988. "Processor Demand and Price-Markings Functions for Catfish: A Disaggregated Analysis with Implications for the Off-Flavor Problems." *Southern Journal of Agricultural Economics*. 20,2:81-91.

Mead, C. 1990. "Experts Agree Aquaculture to Evolve in 1990s, Regulation and Environmental Concerns to Grow." *Catfish News*. 4,6:3.

Mead, J.W. 1989. *Aquaculture Management*. New York: Van Nostrand Reinhold.

McGoodwin, J.R. 1982. "Aquaculture Development in Atomistic Societies," in *Aquaculture Development in Less Developed Countries*, Leah J. Smith and Susan Peterson, eds. Boulder: Westview Press, pp. 61-76.

Shang, Y.C. 1990. "Socioeconomic Constraints of Aquaculture in Asia," *World Aquaculture*. 21,1:34.

Williams, R.R. and D.V. Lightner. 1988. "Regulatory Status of Therapeutants for Penaeid Shrimp Culture in the United States." *Journal of the World Aquaculture Society*. 19,4(1988):188-96.

Appendix

The Consumer Price Index

Market Basket		Base Year (1980)		Current Year (1987)	
Good (1)	Quantity (2)	Price (3)	Expenditure (4)	Price $ (5)	Expenditure (6)
Catfish	76,842	.70	53,789	.61	46,874
Salmon	7,616	.45	3,427	.46	3,503
Trout	48,141	.78	37,550	1.10	52,955
			94,766		103,332

$$P*I=\frac{\sum_{i=1}^{n} P_i^1 * X_i^o}{\sum_{i=o}^{n} P_i^o * X_i^o} *100$$

where:

Pi = price index

$\sum_{i=1}^{n} P_i^o * X_i^o$ = cost of the market basket in the base year

$\sum_{i=1}^{n} P_i^1 * X_i^o$ = cost of the market in the year under investigation

x = goods and services

P = prices

subscript i = identifies goods and services over which the summation is made.

superscript o = relative to base year

superscript 1 = relative to the year in question.

$$PI=\frac{\sum_{i=1}^{n} P_i^1 * X_i^o}{\sum_{i=1} P_i^o * X_i^o} *100=\frac{103,332}{94,766} *100=1.09$$

Thus, the price index has risen 9%. An index is a ratio of two numbers expressed as a %.

The price index measures the cost of a given combination of goods and services in one year as compared to some earlier "base" year. The combination of goods and services is called a "market basket" and reflects the country's average consumption patterns. The relevant question is in connection to th price index is how much do I have to pay this year for a given market basket tha costs $100.00 last year or some other proceeding base year.

In the U. S., the department of Labor market basket includes approximately 400 different goods and services and is revised particularly to bring it up to date. Effort is made to take into consideration changes in quality.

There might be several consumer price indices such as Food, Housing Apparel and upkeep, Transportation, Medical care, Personal care, Reading and Recreation, and all other goods.

To calculate a cost-of-living index number is a very complicated, costly and difficult task involving many technical decisions: such an index number must take into consideration all the important goods and services bought by thousands of customers in many sites.

How index numbers are used.

1. Measure changes that have taken place from one time period to another.

2. Combine changes in several series.

3. Devalue a time series in terms of constant dollars.

The purchasing power of the dollar is constantly changing. Over the short run these changes may be small, but over a long period of time they can be substantial. If a series such as wage rates is divided by its equivalent "cost of living" index for each period, the resulting series is said to be expressed in constant dollars.

Laspeyres' price index

Laspeyres' price index, 1987 (using base-year weights) $P_L = \frac{\sum P_i Q_o}{\sum P_o Q_o} * 100$

Product	P_o	P_{87}	Q_o	P_oQ_o	$P_{87}Qo$
Catfish	.70	.61	76,842	53,789	46,874
Salmon	.45	.46	7,616	3,427	3,503
Trout	.781	.10	48,141	37,550	52,955
				94,766	103,332

$$P_L = \frac{\sum P_{87} Q_o}{\sum P_o Q_o} * 100 = \frac{103,332}{94,766} * 100 = 1.09$$

Paasche's Price Index

$$P_p = \frac{\sum P_i Q_i}{\sum P_o Q_i} * 100$$

The index corresponds to the harmonic mean of price relatives weighted with a given-year values the fact that formula requires that new weights be found.

Paasche's Price Index 1972
(1980 = 100.00)

Product	q_{87}	P_o	P_{87}	$P_{87}q_{87}$	P_oq_{87}
Catfish	375,000	.70	.61	228,750	262,500
Salmon	80,000	.45	.46	36,800	36,000
Trout	59,000	.78	1.10	64,900	46,020
				330,450	344,520

$$P_p = \frac{\sum P_i Q_i}{\sum P_o Q_i} * 100 = \frac{330,450}{344,520} * 100 = 95.9$$

APPENDIX TABLE I. Future Value of $1.00
Vn = $1 (1 + i)n

n	5%	6%	7%	8%	9%	10%	11%	12%	13%
1	1.050	1.060	1.070	1.080	1.090	1.100	1.110	1.120	1.130
2	1.103	1.124	1.145	1.166	1.188	1.210	1.232	1.254	1.277
3	1.158	1.191	1.225	1.260	1.295	1.331	1.368	1.405	1.443
4	1.216	1.262	1.311	1.360	1.412	1.464	1.518	1.574	1.630
5	1.276	1.338	1.403	1.469	1.539	1.611	1.685	1.762	1.842
6	1.340	1.419	1.501	1.587	1.677	1.772	1.870	1.974	2.082
7	1.407	1.504	1.606	1.714	1.828	1.949	2.076	2.211	2.353
8	1.477	1.594	1.718	1.851	1.993	2.144	2.305	2.476	2.658
9	1.551	1.689	1.838	1.999	2.172	2.358	2.558	2.773	3.004
10	1.629	1.791	1.967	2.159	2.367	2.594	2.839	3.106	3.395
11	1.710	1.898	2.105	2.332	2.580	2.853	3.152	3.479	3.836
12	1.796	2.012	2.252	2.518	2.813	3.138	3.498	3.896	4.335
13	1.886	2.133	2.410	2.720	3.066	3.452	3.883	4.363	4.898
14	1.980	2.261	2.579	2.937	3.342	3.797	4.310	4.887	5.535
15	2.079	2.397	2.759	3.172	3.642	4.177	4.785	5.474	6.254
16	2.183	2.540	2.952	3.426	3.970	4.595	5.311	6.130	7.067
17	2.292	2.693	3.159	3.700	4.328	5.054	5.895	6.866	7.986
18	2.407	2.854	3.380	3.996	4.717	5.560	6.544	7.690	9.024
19	2.527	3.026	3.617	4.316	5.142	6.116	7.263	8.613	10.197
20	2.653	3.207	3.870	4.661	5.604	6.727	8.062	9.646	11.523
21	2.786	3.400	4.141	5.034	6.109	7.400	8.949	10.804	13.021
22	2.925	3.604	4.430	5.437	6.659	8.140	9.934	12.100	14.714
23	3.072	3.820	4.741	5.871	7.258	8.954	11.026	13.552	16.627
24	3.225	4.049	5.072	6.341	7.911	9.850	12.239	15.179	18.788
30	4.322	5.743	7.612	10.063	13.268	17.449	22.892	29.960	39.116
36	5.792	8.147	11.424	15.968	22.251	30.913	42.818	59.136	81.437
40	7.040	10.286	14.974	21.725	31.409	45.259	65.001	93.051	132.782
48	10.401	16.394	25.729	40.211	62.585	97.017	149.797	230.391	352.992
50	11.467	18.420	29.457	46.902	74.358	117.391	184.565	289.002	450.736
60	18.679	32.988	57.946	101.257	176.031	304.482	524.057	897.597	1530.053

n	14%	15%	16%	17%	18%	19%	20%	21%	22%
1	1.140	1.150	1.160	1.170	1.180	1.190	1.200	1.210	1.220
2	1.300	1.323	1.346	1.369	1.392	1.416	1.440	1.464	1.488
3	1.482	1.521	1.561	1.602	1.643	1.685	1.728	1.772	1.816
4	1.689	1.749	1.811	1.874	1.939	2.005	2.074	2.144	2.215
5	1.925	2.011	2.100	2.192	2.288	2.386	2.488	2.594	2.703
6	2.195	2.313	2.436	2.565	2.700	2.840	2.986	3.138	3.297
7	2.502	2.660	2.826	3.001	3.185	3.379	3.583	3.797	4.023
8	2.853	3.059	3.278	3.511	3.759	4.021	4.300	4.595	4.908
9	3.252	3.518	3.803	4.108	4.435	4.785	5.160	5.560	5.987
10	3.707	4.046	4.411	4.807	5.234	5.695	6.192	6.727	7.305
11	4.226	4.652	5.117	5.624	6.176	6.777	7.430	8.140	8.912
12	4.818	5.350	5.936	6.580	7.288	8.064	8.916	9.850	10.872
13	5.492	6.153	6.886	7.699	8.599	9.596	10.699	11.918	13.264
14	6.261	7.076	7.988	9.007	10.147	11.420	12.839	14.421	16.182
15	7.138	8.137	9.266	10.539	11.974	13.590	15.407	17.449	19.742
16	8.137	9.358	10.748	12.330	14.129	16.172	18.488	21.114	24.086
17	9.276	10.761	12.468	14.426	16.672	19.244	22.186	25.548	29.384
18	10.575	12.375	14.463	16.879	19.673	22.901	26.623	30.913	35.849
19	12.056	14.232	16.777	19.748	23.214	27.252	31.948	37.404	43.736
20	13.743	16.367	19.461	23.106	27.393	32.429	38.338	45.259	53.358
21	15.668	18.822	22.574	27.034	32.324	38.591	46.005	54.764	65.096
22	17.861	21.645	26.186	31.629	38.142	45.923	55.206	66.264	79.418
23	20.362	24.891	30.376	37.006	45.008	54.649	66.247	80.180	96.889
24	23.212	28.625	35.236	43.297	53.109	65.032	79.497	97.017	118.205
30	50.950	66.212	85.850	111.065	143.371	184.675	237.376	304.482	389.758
36	111.834	153.152	209.164	284.899	387.037	524.434	708.802	955.594	1285.150
40	188.884	267.864	378.721	533.869	750.378	1051.668	1469.772	2048.400	2847.038
48	538.807	819.401	1241.605	1874.655	2820.567	4229.160	6319.749	9412.344	13972.428
50	700.233	1083.657	1670.704	2566.215	3927.357	5988.914	9100.438	13780.612	20796.561
60	2595.919	4383.999	7370.201	12335.356	20555.140	34104.971	56347.514	92709.069	151911.216

APPENDIX TABLE I (continued)

n	23%	24%	25%	26%	27%	28%	29%	30%
1	1.230	1.240	1.250	1.260	1.270	1.280	1.290	1.300
2	1.513	1.538	1.563	1.588	1.613	1.638	1.664	1.690
3	1.861	1.907	1.953	2.000	2.048	2.097	2.147	2.197
4	2.289	2.364	2.441	2.520	2.601	2.684	2.769	2.856
5	2.815	2.932	3.052	3.176	3.304	3.436	3.572	3.713
6	3.463	3.635	3.815	4.002	4.196	4.398	4.608	4.827
7	4.259	4.508	4.768	5.042	5.329	5.629	5.945	6.275
8	5.239	5.590	5.960	6.353	6.768	7.206	7.669	8.157
9	6.444	6.931	7.451	8.005	8.595	9.223	9.893	10.604
10	7.926	8.594	9.313	10.086	10.915	11.806	12.761	13.786
11	9.749	10.657	11.642	12.708	13.862	15.112	16.462	17.922
12	11.991	13.215	14.552	16.012	17.605	19.343	21.236	23.298
13	14.749	16.386	18.190	20.175	22.359	24.759	27.395	30.288
14	18.141	20.319	22.737	25.421	28.396	31.691	35.339	39.374
15	22.314	25.196	28.422	32.030	36.062	40.565	45.587	51.186
16	27.446	31.243	35.527	40.358	45.799	51.923	58.808	66.542
17	33.759	38.741	44.409	50.851	58.165	66.461	75.862	86.504
18	41.523	48.039	55.511	64.072	73.870	85.071	97.862	112.455
19	51.074	59.568	69.389	80.731	93.815	108.890	126.242	146.192
20	62.821	73.864	86.736	101.721	119.145	139.380	162.852	190.050
21	77.269	91.592	108.420	128.169	151.314	178.406	210.080	247.065
22	95.041	113.574	135.525	161.492	192.168	228.360	271.003	321.184
23	116.901	140.831	169.407	203.480	244.054	292.300	349.593	417.539
24	143.788	174.631	211.758	256.385	309.948	374.144	450.976	542.801
30	497.913	634.820	807.794	1025.927	1300.504	1645.505	2078.219	2619.996
36	1724.186	2307.707	3081.488	4105.250	5456.749	7237.006	9577.002	12646.219
40	3946.430	5455.913	7523.164	10347.175	14195.439	19426.689	26520.909	36118.865
48	20674.992	30495.860	44841.551	65733.410	96067.968	139984.046	203378.994	294632.676
50	31279.195	46890.435	70064.923	104358.362	154948.026	229349.862	338442.984	497929.223
60	247917.216	402996.347	652530.447	1052525.695	1691310.158	2707685.248	4318994.171	6864377.173

APPENDIX TABLE II. Present Value of $1.00,
Vn = $1(1 + i) − n

n	5%	6%	7%	8%	9%	10%	11%	12%	13%
1	0.952	0.943	0.935	0.926	0.917	0.909	0.901	0.893	0.885
2	0.907	0.890	0.873	0.857	0.842	0.826	0.812	0.797	0.783
3	0.864	0.840	0.816	0.794	0.772	0.751	0.731	0.712	0.693
4	0.823	0.792	0.763	0.735	0.708	0.683	0.659	0.636	0.613
5	0.784	0.747	0.713	0.681	0.650	0.621	0.593	0.567	0.543
6	0.746	0.705	0.666	0.630	0.596	0.564	0.535	0.507	0.480
7	0.711	0.665	0.623	0.583	0.547	0.513	0.482	0.452	0.425
8	0.677	0.627	0.582	0.540	0.502	0.467	0.434	0.404	0.376
9	0.645	0.592	0.544	0.500	0.460	0.424	0.391	0.361	0.333
10	0.614	0.558	0.508	0.463	0.422	0.386	0.352	0.322	0.295
11	0.585	0.527	0.475	0.429	0.388	0.350	0.317	0.287	0.261
12	0.557	0.497	0.444	0.397	0.356	0.319	0.286	0.257	0.231
13	0.530	0.469	0.415	0.368	0.326	0.290	0.258	0.229	0.204
14	0.505	0.442	0.388	0.340	0.299	0.263	0.232	0.205	0.181
15	0.481	0.417	0.362	0.315	0.275	0.239	0.209	0.183	0.160
16	0.458	0.394	0.339	0.292	0.252	0.218	0.188	0.163	0.141
17	0.436	0.371	0.317	0.270	0.231	0.198	0.170	0.146	0.125
18	0.416	0.350	0.296	0.250	0.212	0.180	0.153	0.130	0.111
19	0.396	0.331	0.277	0.232	0.194	0.164	0.138	0.116	0.098
20	0.377	0.312	0.258	0.215	0.178	0.149	0.124	0.104	0.087
21	0.359	0.294	0.242	0.199	0.164	0.135	0.112	0.093	0.077
22	0.342	0.278	0.226	0.184	0.150	0.123	0.101	0.083	0.068
23	0.326	0.262	0.211	0.170	0.138	0.112	0.091	0.074	0.060
24	0.310	0.247	0.197	0.158	0.126	0.102	0.082	0.066	0.053
30	0.231	0.174	0.131	0.099	0.075	0.057	0.044	0.033	0.026
36	0.173	0.123	0.088	0.063	0.045	0.032	0.023	0.017	0.012
40	0.142	0.097	0.067	0.046	0.032	0.022	0.015	0.011	0.008
48	0.096	0.061	0.039	0.025	0.016	0.010	0.007	0.004	0.003
50	0.087	0.054	0.034	0.021	0.013	0.009	0.005	0.003	0.002
60	0.054	0.030	0.017	0.010	0.006	0.003	0.002	0.001	0.001

APPENDIX TABLE II (continued)

n	14%	15%	16%	17%	18%	19%	20%	21%	22%
1	0.877	0.870	0.862	0.855	0.847	0.840	0.833	0.826	0.820
2	0.769	0.756	0.743	0.731	0.718	0.706	0.694	0.683	0.672
3	0.675	0.658	0.641	0.624	0.609	0.593	0.579	0.564	0.551
4	0.592	0.572	0.552	0.534	0.516	0.499	0.482	0.467	0.451
5	0.519	0.497	0.476	0.456	0.437	0.419	0.402	0.386	0.370
6	0.456	0.432	0.410	0.390	0.370	0.352	0.335	0.319	0.303
7	0.400	0.376	0.354	0.333	0.314	0.296	0.279	0.263	0.249
8	0.351	0.327	0.305	0.285	0.266	0.249	0.233	0.218	0.204
9	0.308	0.284	0.263	0.243	0.225	0.209	0.194	0.180	0.167
10	0.270	0.247	0.227	0.208	0.191	0.176	0.162	0.149	0.137
11	0.237	0.215	0.195	0.178	0.162	0.148	0.135	0.123	0.112
12	0.208	0.187	0.168	0.152	0.137	0.124	0.112	0.102	0.092
13	0.182	0.163	0.145	0.130	0.116	0.104	0.093	0.084	0.075
14	0.160	0.141	0.125	0.111	0.099	0.088	0.078	0.069	0.062
15	0.140	0.123	0.108	0.095	0.084	0.074	0.065	0.057	0.051
16	0.123	0.107	0.093	0.081	0.071	0.062	0.054	0.047	0.042
17	0.108	0.093	0.080	0.069	0.060	0.052	0.045	0.039	0.034
18	0.095	0.081	0.069	0.059	0.051	0.044	0.038	0.032	0.028
19	0.083	0.070	0.060	0.051	0.043	0.037	0.031	0.027	0.023
20	0.073	0.061	0.051	0.043	0.037	0.031	0.026	0.022	0.019
21	0.064	0.053	0.044	0.037	0.031	0.026	0.022	0.018	0.015
22	0.056	0.046	0.038	0.032	0.026	0.022	0.018	0.015	0.013
23	0.049	0.040	0.033	0.027	0.022	0.018	0.015	0.012	0.010
24	0.043	0.035	0.028	0.023	0.019	0.015	0.013	0.010	0.008
30	0.020	0.015	0.012	0.009	0.007	0.005	0.004	0.003	0.003
36	0.009	0.007	0.005	0.004	0.003	0.002	0.001	0.001	0.001
40	0.005	0.004	0.003	0.002	0.001	0.001	0.001	0.000	0.000
48	0.002	0.001	0.001	0.001	0.000	0.000	0.000	0.000	0.000
50	0.001	0.001	0.001	0.000	0.000	0.000	0.000	0.000	0.000
60	0.000	0.000	0.000	0.000	0.000	0.000	0.000	0.000	0.000

n	23%	24%	25%	26%	27%	28%	29%	30%	31%	32%
1	0.813	0.806	0.800	0.794	0.787	0.781	0.775	0.769	0.763	0.758
2	0.661	0.650	0.640	0.630	0.620	0.610	0.601	0.592	0.583	0.574
3	0.537	0.524	0.512	0.500	0.488	0.477	0.466	0.455	0.445	0.435
4	0.437	0.423	0.410	0.397	0.384	0.373	0.361	0.350	0.340	0.329
5	0.355	0.341	0.328	0.315	0.303	0.291	0.280	0.269	0.259	0.250
6	0.289	0.275	0.262	0.250	0.238	0.227	0.217	0.207	0.198	0.189
7	0.235	0.222	0.210	0.198	0.188	0.178	0.168	0.159	0.151	0.143
8	0.191	0.179	0.168	0.157	0.148	0.139	0.130	0.123	0.115	0.108
9	0.155	0.144	0.134	0.125	0.116	0.108	0.101	0.094	0.088	0.082
10	0.126	0.116	0.107	0.099	0.092	0.085	0.078	0.073	0.067	0.062
11	0.103	0.094	0.086	0.079	0.072	0.066	0.061	0.056	0.051	0.047
12	0.083	0.076	0.069	0.062	0.057	0.052	0.047	0.043	0.039	0.036
13	0.068	0.061	0.055	0.050	0.045	0.040	0.037	0.033	0.030	0.027
14	0.055	0.049	0.044	0.039	0.035	0.032	0.028	0.025	0.023	0.021
15	0.045	0.040	0.035	0.031	0.028	0.025	0.022	0.020	0.017	0.016
16	0.036	0.032	0.028	0.025	0.022	0.019	0.017	0.015	0.013	0.012
17	0.030	0.026	0.023	0.020	0.017	0.015	0.013	0.012	0.010	0.009
18	0.024	0.021	0.018	0.016	0.014	0.012	0.010	0.009	0.008	0.007
19	0.020	0.017	0.014	0.012	0.011	0.009	0.008	0.007	0.006	0.005
20	0.016	0.014	0.012	0.010	0.008	0.007	0.006	0.005	0.005	0.004
21	0.013	0.011	0.009	0.008	0.007	0.006	0.005	0.004	0.003	0.003
22	0.011	0.009	0.007	0.006	0.005	0.004	0.004	0.003	0.003	0.002
23	0.009	0.007	0.006	0.005	0.004	0.003	0.003	0.002	0.002	0.002
24	0.007	0.006	0.005	0.004	0.003	0.003	0.002	0.002	0.002	0.001
30	0.002	0.002	0.001	0.001	0.001	0.001	0.000	0.000	0.000	0.000
36	0.001	0.000	0.000	0.000	0.000	0.000	0.000	0.000	0.000	0.000
40	0.000	0.000	0.000	0.000	0.000	0.000	0.000	0.000	0.000	0.000
48	0.000	0.000	0.000	0.000	0.000	0.000	0.000	0.000	0.000	0.000
50	0.000	0.000	0.000	0.000	0.000	0.000	0.000	0.000	0.000	0.000
60	0.000	0.000	0.000	0.000	0.000	0.000	0.000	0.000	0.000	0.000

APPENDIX TABLE III. Future Value of a Uniform Series, $V_n = \$1(((1 + i)^n - 1)/i)$

n	5%	6%	7%	8%	9%	10%	11%	12%	13%
1	1.000	1.000	1.000	1.000	1.000	1.000	1.000	1.000	1.000
2	2.050	2.060	2.070	2.080	2.090	2.100	2.110	2.120	2.130
3	3.153	3.184	3.215	3.246	3.278	3.310	3.342	3.374	3.407
4	4.310	4.375	4.440	4.506	4.573	4.641	4.710	4.779	4.850
5	5.526	5.637	5.751	5.867	5.985	6.105	6.228	6.353	6.480
6	6.802	6.975	7.153	7.336	7.523	7.716	7.913	8.115	8.323
7	8.142	8.394	8.654	8.923	9.200	9.487	9.783	10.089	10.405
8	9.549	9.897	10.260	10.637	11.028	11.436	11.859	12.300	12.757
9	11.027	11.491	11.978	12.488	13.021	13.579	14.164	14.776	15.416
10	12.578	13.181	13.816	14.487	15.193	15.937	16.722	17.549	18.420
11	14.207	14.972	15.784	16.645	17.560	18.531	19.561	20.655	21.814
12	15.917	16.870	17.888	18.977	20.141	21.384	22.713	24.133	25.650
13	17.713	18.882	20.141	21.495	22.953	24.523	26.212	28.029	29.985
14	19.599	21.015	22.550	24.215	26.019	27.975	30.095	32.393	34.883
15	21.579	23.276	25.129	27.152	29.361	31.772	34.405	37.280	40.417
16	23.657	25.673	27.888	30.324	33.003	35.950	39.190	42.753	46.672
17	25.840	28.213	30.840	33.750	36.974	40.545	44.501	48.884	53.739
18	28.132	30.906	33.999	37.450	41.301	45.599	50.396	55.750	61.725
19	30.539	33.760	37.379	41.446	46.018	51.159	56.939	63.440	70.749
20	33.066	36.786	40.995	45.762	51.160	57.275	64.203	72.052	80.947
21	35.719	39.993	44.865	50.423	56.765	64.002	72.265	81.699	92.470
22	38.505	43.392	49.006	55.457	62.873	71.403	81.214	92.503	105.491
23	41.430	46.996	53.436	60.893	69.532	79.543	91.148	104.603	120.205
24	44.502	50.816	58.177	66.765	76.790	88.497	102.174	118.155	136.831
30	66.439	79.058	94.461	113.283	136.308	164.494	199.021	241.333	293.199
36	95.836	119.121	148.913	187.102	236.125	299.127	380.164	484.463	618.749
40	120.800	154.762	199.635	259.057	337.882	442.593	581.826	767.091	1013.704
48	188.025	256.565	353.270	490.132	684.280	960.172	1352.700	1911.590	2707.633
50	209.348	290.336	406.529	573.770	815.084	1163.909	1668.771	2400.018	3459.507
60	353.584	533.128	813.520	1253.213	1944.792	3034.816	4755.066	7471.641	11761.950

n	14%	15%	16%	17%	18%	19%	20%	21%	22%
1	1.000	1.000	1.000	1.000	1.000	1.000	1.000	1.000	1.000
2	2.140	2.150	2.160	2.170	2.180	2.190	2.200	2.210	2.220
3	3.440	3.473	3.506	3.539	3.572	3.606	3.640	3.674	3.708
4	4.921	4.993	5.066	5.141	5.215	5.291	5.368	5.446	5.524
5	6.610	6.742	6.877	7.014	7.154	7.297	7.442	7.589	7.740
6	8.536	8.754	8.977	9.207	9.442	9.683	9.930	10.183	10.442
7	10.730	11.067	11.414	11.772	12.142	12.523	12.916	13.321	13.740
8	13.233	13.727	14.240	14.773	15.327	15.902	16.499	17.119	17.762
9	16.085	16.786	17.519	18.285	19.086	19.923	20.799	21.714	22.670
10	19.337	20.304	21.321	22.393	23.521	24.709	25.959	27.274	28.657
11	23.045	24.349	25.733	27.200	28.755	30.404	32.150	34.001	35.962
12	27.271	29.002	30.850	32.824	34.931	37.180	39.581	42.142	44.874
13	32.089	34.352	36.786	39.404	42.219	45.244	48.497	51.991	55.746
14	37.581	40.505	43.672	47.103	50.818	54.841	59.196	63.909	69.010
15	43.842	47.580	51.660	56.110	60.965	66.261	72.035	78.330	85.192
16	50.980	55.717	60.925	66.649	72.939	79.850	87.442	95.780	104.935
17	59.118	65.075	71.673	78.979	87.068	96.022	105.931	116.894	129.020
18	68.394	75.836	84.141	93.406	103.740	115.266	128.117	142.441	158.405
19	78.969	88.212	98.603	110.285	123.414	138.166	154.740	173.354	194.254
20	91.025	102.444	115.380	130.033	146.628	165.418	186.688	210.758	237.989
21	104.768	118.810	134.841	153.139	174.021	197.847	225.026	256.018	291.347
22	120.436	137.632	157.415	180.172	206.345	236.438	271.031	310.781	356.443
23	138.297	159.276	183.601	211.801	244.487	282.362	326.237	377.045	435.861
24	158.659	184.168	213.978	248.808	289.494	337.010	392.484	457.225	532.750
30	356.787	434.745	530.312	647.439	790.948	966.712	1181.882	1445.151	1767.081
36	791.673	1014.346	1301.027	1669.994	2144.649	2754.914	3539.009	4545.685	5837.047
40	1342.025	1779.090	2360.757	3134.522	4163.213	5529.829	7343.858	9749.525	12936.535
48	3841.475	5456.005	7753.782	11021.500	15664.259	22253.475	31593.744	44815.922	63506.490
50	4994.521	7217.716	10435.649	15089.502	21813.094	31515.336	45497.191	65617.202	94525.279
60	18535.133	29219.992	46057.509	72555.038	114189.666	179494.584	281732.572	441466.994	690500.982

APPENDIX TABLE III (continued)

n	23%	24%	25%	26%	27%	28%	29%	30%	31%
1	1.000	1.000	1.000	1.000	1.000	1.000	1.000	1.000	1.000
2	2.230	2.240	2.250	2.260	2.270	2.280	2.290	2.300	2.310
3	3.743	3.778	3.813	3.848	3.883	3.918	3.954	3.990	4.026
4	5.604	5.684	5.766	5.848	5.931	6.016	6.101	6.187	6.274
5	7.893	8.048	8.207	8.368	8.533	8.700	8.870	9.043	9.219
6	10.708	10.980	11.259	11.544	11.837	12.136	12.442	12.756	13.077
7	14.171	14.615	15.073	15.546	16.032	16.534	17.051	17.583	18.131
8	18.430	19.123	19.842	20.588	21.361	22.163	22.995	23.858	24.752
9	23.669	24.712	25.802	26.940	28.129	29.369	30.664	32.015	33.425
10	30.113	31.643	33.253	34.945	36.723	38.593	40.556	42.619	44.786
11	38.039	40.238	42.566	45.031	47.639	50.398	53.318	56.405	59.670
12	47.788	50.895	54.208	57.739	61.501	65.510	69.780	74.327	79.168
13	59.779	64.110	68.760	73.751	79.107	84.853	91.016	97.625	104.710
14	74.528	80.496	86.949	93.926	101.465	109.612	118.411	127.913	138.170
15	92.669	100.815	109.687	119.347	129.861	141.303	153.750	167.286	182.003
16	114.983	126.011	138.109	151.377	165.924	181.868	199.337	218.472	239.423
17	142.430	157.253	173.636	191.735	211.723	233.791	258.145	285.014	314.645
18	176.188	195.994	218.045	242.585	269.888	300.252	334.007	371.518	413.185
19	217.712	244.033	273.556	306.658	343.758	385.323	431.870	483.973	542.272
20	268.785	303.601	342.945	387.389	437.573	494.213	558.112	630.165	711.376
21	331.606	377.465	429.681	489.110	556.717	633.593	720.964	820.215	932.903
22	408.875	469.056	538.101	617.278	708.031	811.999	931.044	1067.280	1223.103
23	503.917	582.630	673.626	778.771	900.199	1040.358	1202.047	1388.464	1603.264
24	620.817	723.461	843.033	982.251	1144.253	1332.659	1551.640	1806.003	2101.276
30	2160.491	2640.916	3227.174	3942.026	4812.977	5873.231	7162.824	8729.985	10632.746
36	7492.111	9611.279	12321.952	15785.577	20206.477	25842.877	33020.696	42150.729	53750.052
40	17154.046	22728.803	30088.655	39792.982	52571.998	69377.460	91447.963	120392.883	158300.134
48	89886.922	127061.917	179362.203	252816.963	355803.586	499939.451	701303.427	982105.588	1372965.035
50	135992.154	195372.644	280255.693	401374.471	573877.874	819103.077	1167041.323	1659760.743	2356147.606
60	1077896.591	1679147.280	2610117.787	4048171.905	6264107.994	9670300.886	14893079.901	22881253.909	35068404.358

APPENDIX TABLE IV. Present Value of a Uniform Series,
$V_n = \$1((1-(1/(1+i)n))/i)$

n	5%	6%	7%	8%	9%	10%	11%	12%	13%
1	0.952	0.943	0.935	0.926	0.917	0.909	0.901	0.893	0.885
2	1.859	1.833	1.808	1.783	1.759	1.736	1.713	1.690	1.668
3	2.723	2.673	2.624	2.577	2.531	2.487	2.444	2.402	2.361
4	3.546	3.465	3.387	3.312	3.240	3.170	3.102	3.037	2.974
5	4.329	4.212	4.100	3.993	3.890	3.791	3.696	3.605	3.517
6	5.076	4.917	4.767	4.623	4.486	4.355	4.231	4.111	3.998
7	5.786	5.582	5.389	5.206	5.033	4.868	4.712	4.564	4.423
8	6.463	6.210	5.971	5.747	5.535	5.335	5.146	4.968	4.799
9	7.108	6.802	6.515	6.247	5.995	5.759	5.537	5.328	5.132
10	7.722	7.360	7.024	6.710	6.418	6.145	5.889	5.650	5.426
11	8.306	7.887	7.499	7.139	6.805	6.495	6.207	5.938	5.687
12	8.863	8.384	7.943	7.536	7.161	6.814	6.492	6.194	5.918
13	9.394	8.853	8.358	7.904	7.487	7.103	6.750	6.424	6.122
14	9.899	9.295	8.745	8.244	7.786	7.367	6.982	6.628	6.302
15	10.380	9.712	9.108	8.559	8.061	7.606	7.191	6.811	6.462
16	10.838	10.106	9.447	8.851	8.313	7.824	7.379	6.974	6.604
17	11.274	10.477	9.763	9.122	8.544	8.022	7.549	7.120	6.729
18	11.690	10.828	10.059	9.372	8.756	8.201	7.702	7.250	6.840
19	12.085	11.158	10.336	9.604	8.950	8.365	7.839	7.366	6.938
20	12.462	11.470	10.594	9.818	9.129	8.514	7.963	7.469	7.025
21	12.821	11.764	10.836	10.017	9.292	8.649	8.075	7.562	7.102
22	13.163	12.042	11.061	10.201	9.442	8.772	8.176	7.645	7.170
23	13.489	12.303	11.272	10.371	9.580	8.883	8.266	7.718	7.230
24	13.799	12.550	11.469	10.529	9.707	8.985	8.348	7.784	7.283
30	15.372	13.765	12.409	11.258	10.274	9.427	8.694	8.055	7.496
36	16.547	14.621	13.035	11.717	10.612	9.677	8.879	8.192	7.598
40	17.159	15.046	13.332	11.925	10.757	9.779	8.951	8.244	7.634
48	18.077	15.650	13.730	12.189	10.934	9.897	9.030	8.297	7.671
50	18.256	15.762	13.801	12.233	10.962	9.915	9.042	8.304	7.675
60	18.929	16.161	14.039	12.377	11.048	9.967	9.074	8.324	7.687

APPENDIX TABLE IV (continued)

n	14%	15%	16%	17%	18%	19%	20%	21%	22%
1	0.877	0.870	0.862	0.855	0.847	0.840	0.833	0.826	0.820
2	1.647	1.626	1.605	1.585	1.566	1.547	1.528	1.509	1.492
3	2.322	2.283	2.246	2.210	2.174	2.140	2.106	2.074	2.042
4	2.914	2.855	2.798	2.743	2.690	2.639	2.589	2.540	2.494
5	3.433	3.352	3.274	3.199	3.127	3.058	2.991	2.926	2.864
6	3.889	3.784	3.685	3.589	3.498	3.410	3.326	3.245	3.167
7	4.288	4.160	4.039	3.922	3.812	3.706	3.605	3.508	3.416
8	4.639	4.487	4.344	4.207	4.078	3.954	3.837	3.726	3.619
9	4.946	4.772	4.607	4.451	4.303	4.163	4.031	3.905	3.786
10	5.216	5.019	4.833	4.659	4.494	4.339	4.192	4.054	3.923
11	5.453	5.234	5.029	4.836	4.656	4.486	4.327	4.177	4.035
12	5.660	5.421	5.197	4.988	4.793	4.611	4.439	4.278	4.127
13	5.842	5.583	5.342	5.118	4.910	4.715	4.533	4.362	4.203
14	6.002	5.724	5.468	5.229	5.008	4.802	4.611	4.432	4.265
15	6.142	5.847	5.575	5.324	5.092	4.876	4.675	4.489	4.315
16	6.265	5.954	5.668	5.405	5.162	4.938	4.730	4.536	4.357
17	6.373	6.047	5.749	5.475	5.222	4.990	4.775	4.576	4.391
18	6.467	6.128	5.818	5.534	5.273	5.033	4.812	4.608	4.419
19	6.550	6.198	5.877	5.584	5.316	5.070	4.843	4.635	4.442
20	6.623	6.259	5.929	5.628	5.353	5.101	4.870	4.657	4.460
21	6.687	6.312	5.973	5.665	5.384	5.127	4.891	4.675	4.476
22	6.743	6.359	6.011	5.696	5.410	5.149	4.909	4.690	4.488
23	6.792	6.399	6.044	5.723	5.432	5.167	4.925	4.703	4.499
24	6.835	6.434	6.073	5.746	5.451	5.182	4.937	4.713	4.507
30	7.003	6.566	6.177	5.829	5.517	5.235	4.979	4.746	4.534
36	7.079	6.623	6.220	5.862	5.541	5.253	4.993	4.757	4.542
40	7.105	6.642	6.233	5.871	5.548	5.258	4.997	4.760	4.544
48	7.130	6.659	6.245	5.879	5.554	5.262	4.999	4.761	4.545
50	7.133	6.661	6.246	5.880	5.554	5.262	4.999	4.762	4.545
60	7.140	6.665	6.249	5.882	5.555	5.263	5.000	4.762	4.545

n	23%	24%	25%	26%	27%	28%	29%	30%	31%
1	0.813	0.806	0.800	0.794	0.787	0.781	0.775	0.769	0.763
2	1.474	1.457	1.440	1.424	1.407	1.392	1.376	1.361	1.346
3	2.011	1.981	1.952	1.923	1.896	1.868	1.842	1.816	1.791
4	2.448	2.404	2.362	2.320	2.280	2.241	2.203	2.166	2.130
5	2.803	2.745	2.689	2.635	2.583	2.532	2.483	2.436	2.390
6	3.092	3.020	2.951	2.885	2.821	2.759	2.700	2.643	2.588
7	3.327	3.242	3.161	3.083	3.009	2.937	2.868	2.802	2.739
8	3.518	3.421	3.329	3.241	3.156	3.076	2.999	2.925	2.854
9	3.673	3.566	3.463	3.366	3.273	3.184	3.100	3.019	2.942
10	3.799	3.682	3.571	3.465	3.364	3.269	3.178	3.092	3.009
11	3.902	3.776	3.656	3.543	3.437	3.335	3.239	3.147	3.060
12	3.985	3.851	3.725	3.606	3.493	3.387	3.286	3.190	3.100
13	4.053	3.912	3.780	3.656	3.538	3.427	3.322	3.223	3.129
14	4.108	3.962	3.824	3.695	3.573	3.459	3.351	3.249	3.152
15	4.153	4.001	3.859	3.726	3.601	3.483	3.373	3.268	3.170
16	4.189	4.033	3.887	3.751	3.623	3.503	3.390	3.283	3.183
17	4.219	4.059	3.910	3.771	3.640	3.518	3.403	3.295	3.193
18	4.243	4.080	3.928	3.786	3.654	3.529	3.413	3.304	3.201
19	4.263	4.097	3.942	3.799	3.664	3.539	3.421	3.311	3.207
20	4.279	4.110	3.954	3.808	3.673	3.546	3.427	3.316	3.211
21	4.292	4.121	3.963	3.816	3.679	3.551	3.432	3.320	3.215
22	4.302	4.130	3.970	3.822	3.684	3.556	3.436	3.323	3.217
23	4.311	4.137	3.976	3.827	3.689	3.559	3.438	3.325	3.219
24	4.318	4.143	3.981	3.831	3.692	3.562	3.441	3.327	3.221
30	4.339	4.160	3.995	3.842	3.701	3.569	3.447	3.332	3.225
36	4.345	4.165	3.999	3.845	3.703	3.571	3.448	3.333	3.226
40	4.347	4.166	3.999	3.846	3.703	3.571	3.448	3.333	3.226
48	4.348	4.167	4.000	3.846	3.704	3.571	3.448	3.333	3.226
50	4.348	4.167	4.000	3.846	3.704	3.571	3.448	3.333	3.226
60	4.348	4.167	4.000	3.846	3.704	3.571	3.448	3.333	3.226

Author Citations Index

Alsagoff, S.A.K., 133,282,284
Andrews, J.W., 83
Anonymous, 118

Bailey, L.C., 1
Bang Ho, 11
Banks, J.L., 118
Bardach, J.E., 134,148
Bell, F.W., 39
Ben-Yami, M., 16
Brett, J.R., 81,82

Cacho, O.J., 82,118,121,122,128, 130,225
Clonts, H.A., 143,278,286
Cole, B.A., 121
Crews, J.R., 215

Dellenbarger, L.E., 40,41,50
Dupree, H.K., 84

Engle, C.R., 105,225

FAO, 10,11,15,141
Fowler, L.G., 117,118

Hastings, W.H., 84,85
Hebicha, H., 280

Ives, B.H., 278

James, D.G., 1,9
Jensen, J.W., 215
Jolly, C.M., 225

Kamara, A.B., 278
Katoh, J., 144
Kent, G., 3,9
Kinnucan, H., 41,281
Kromhout, D., 11

Liao, I., 12
Lovell, R.T., 12

Marshall, A., 67
Mitchison, A., 3
Miyamura, M., 144
McGoodwin, J.R., 279

Nash, C.E., 6,7
Nerrie, B., 93

Pavelec, C.L., 12,38
Penson, J.B., 228

Rabanal, H.R., 1,3,143
Reutebuch, E.M., 84
Rupp, E.M., 40

Secretan, P.A.D., 221,224,228
Shang, Y.C., 278
Stickney, R.R., 83

USDA, 1,4,226,267

Van Dam, A.A., 93,136

West, B.W., 83
Wineholt, D., 41
Williams, S.B., 143
Williams, R.R., 283

Subject-Author Index

Abella, C., 275
Absolute value, 44,45,55
Acreage control, 287
Adams, C.M., 116,275
Adrian, J.L., 187
Advertising, 261
Africa, 5,6,7,278
Aggregate demand. *See* Market demand
Aiken, D.E., 290
Alabama, 7,93,175,215,225,257, 282
Alaska, 257,280
Albrecht, W.P., Jr., 34,73,95
Aldrich, D.V., 189
Algeria, 7
Allen, G.P., 116,187
Allocation of resources, 61
Al-Meeri, A., 189
Alsagoff, S.A.K., 138,290
Amortization. *See* Depreciation
Anchovy, 118
Anderson, C.M., 275
Andrews, J.W., 97
Animal protein, 1
Aquaculture, 21,61,143,146,257, 278
  advantages of, 12
  growth of, 3
  in economic development, 13
  limitations of, 16
Aquaculture economics, 34
Aquacultural production, 3,92,94, 141,148,166
Arkansas, 7,118
Armstrong, M.S., 73
Arya, J., 34
Asia, 5,6,7,38,61
  Southeast, 12
  West, 7
  East, 5,7
Asset fixity, 283
Assets, 181
  depreciable, 159
Atwood, J., 189
Avault, J.W., Jr., 138
Average cost, 103,251
  cost curves, 104,108,109
  defined, 103,104
  fixed cost, 103-104
  in the long-run, 110-115. *See also* Cost curves
Average physical product, defined 86
  curve, 86. *See also* Production function
  maximum value, 87,89
Average revenue, 54
Average total cost, 103-104,251
Average variable cost, 103-104,108
  curve, 103,109-110

Baer, C., 18
Bailey, C., 17
Baker, C.B., 201,234
Balance sheet, 181
Baldwin, W.J., 190
Bank, 147,193,201
Banks, J.L., 138
Barry, P.J., 201,234
Bauer, L.L., 187,190
Bebee, D., 73
Benefit-cost, 212

Berrigan, M.E., 97
Biological, 83,92
Biotechnology, 282
Bishop, C.E., 138
Blaylock, J.R., 73
Blommestein, E., 116,188
Boehlje, M.D., 201,235
Bosschieter, E., 18
Botsford, L.W., 96,116,190
Boyd, C.E., 73,138,187
Brazil, 272
Break-even analysis, 176
Brick, R.E., 116
Brigham, E.F., 201,235
Brown, J.H., 290
Browning, E.K., 34,116,256
Browning, J.M., 34,116,256
Budget 150,151
  enterprise budgets, 173
Budget analysis, 151,173
Budgeting, 203,221,233
Business, 149,253

Cacho, O., 275
Cameroon. *See* The Cameroon
Campbell, M., 275
Campbell-Asselbergs, E., 17
Canada, 116
Capital budgeting, 203,221,233
Capps, O., 73
Caribbean, 5,6,8
Carp, 4,133-134,146
  grass, 4,134,146
Cartel, 253
Catfish
  biotechnology, 282
  consumed during Lent, 40
  demand curve, 80-85
  example of production, 52
  industry, 5
  marketing, 257,269,272
  production, 80,93,105-108
  risk in stocking, 229
  supply curve, 66
Catholic, 40. *See also* Lent
Cash flow, 205,213,233
Chang, K., 276
Change in demand, 38
Change in price, 60,62,67
Change in supply, 70
Charmantier, G., 290
Charmantier-Daures, M., 290
Chaston, I., 188,275
Chien, Y., 138
China, 3,5,6,278
Choice, 22
Chong, K., 188
Chong, K.C., 96
Clark, C.T., 34
Clark, T.M., 116
Clonts, H.A., 138,290
Collusion. *See* Cartel
Commercial fishing, 3,5
Competition, monopolistic, 250
  perfect, 237
  perfectly competitive markets, 91
  price, 248
  under monopoly, 243
  under oligopoly, 248
Complementarity, 130
Complementary goods, 37
Compound interest, 193
Compounding, 192
Conner, J.R., 256,276
Consumer demand, 35
Consumer price index, 27,291
Consumer behavior, 35
  complementary goods, 37
  demand, 35
  income effect, 41
  inferior goods, 50
  normal goods, 50
  substitute goods, 48
  substitution effect, 48,51
  superior goods, 50
  time preference, 192
Consumers, assumptions underlying decisions of, 48

knowledge pertaining to decisions of, 37
Contracting, 225
Cost curves, 103-115,127
average cost, 103-104
envelope curve, 111
fixed cost, 101
long-run, 109-111
marginal cost, 102-112
simulated for catfish, 105
variable cost, 101
total cost, 101,107
Cost of production, 99
and profit maximization, 106
average, 103
diseconomies of scale, 114
economies of scale, 112
fixed cost, 100
long-run, 110
marginal, 102
planning horizon, 109
total cost, 100
variable cost, 101. See also, Production
Costa Rica, 287
Coulander, C.L., 18
Courtenay, W.R., Jr., 17
Coutts, D.C., 17
Crawfish, 6
Credit, 285
Crews, J., 188
Crop, 83,93
Cross elasticity, 51
Curves. *See* Cost curves
Cycon, D., 17

Darby, M.R., 34
Debt-to-equity ratio, 184
De Castillo, V.G., 189
Decision making, 90
Deese, H., 116,188
Definitions, 24. *See* specific terms
Dehadrai, P.V., 96
Dellenbarger, L.E., 73,96,116,256
Demand, change in, 39
derived, 57
elasticity of, 42-51
factors affecting, 38
law of, 36
monopoly, 245
perfectly competitive market equilibrium, 69-71
theoretical, 37. *See also* Consumer demand; Market demand; Elasticity of demand
Demand curve, 36
and revenue, 56
defined, 36
elasticity and shape of, 43-44,53
for perfectly competitive firm, 237
for monopolist, 245
for oligopoly, 248
market equilibrium, 69. *See also* Elasticity of demand
Demand and supply analysis. *See* Monopoly; Oligopoly; Monopolistic competition
Demand expansion, 286
Demand schedule, 36,53
Denmark, 16
Depreciation, and NNP, 26
declining balance, 167
defined, 166
non-cash expense, 162
straight line, 166
sum of digits, 168
Diet, 42
Dillard, J., 73,256
Discounting, 196,205,208,233
Diminishing marginal rate of technical substitution, 120-121
Diseconomies of scale, 114
Dupree, H.K., 96,138
Durda, J., 188
Dyerberg, J., 17
Dzidzienyo, S., 18

Earnings, 196
Economic activity, circular flow of, 30
Economic development, 13
Economic goods, 21,24
Economic policies, 287
Economic region of production, 132
  ridge lines defining, 132
Economic systems, 28
  mixed, 31
  types of, 30
Economics, defined, 21
  normative, 24
  positive, 24
Economies of scale, 112
Effective demand, 35
Efficiency, relation to average physical product, 87
  relation to marginal cost, 102
  relation to average total cost, 104
Egypt, 7
Elasticity of demand, along demand curve, 53
  related to revenue, 54
  calculation of, 45
  cross-price, 51;
  defined, 42
  income, 49
  price, 42,48,51
  ranges for linear demand curve, 53
  unitary, 55,56. *See also* Demand
Elasticity of supply, 63
  calculation of, 64
  defined, 63
  relation to time, 68. *See also* Supply
El Naggar, G.O., 139
Energy, total, 82
  maintenance, 82
Engel, C., 73,188,189,234,275
Enterprise, analyzing farm, 151
  budget, 173
  measuring income, 152
Envelope curve, for long-run, 110. *See also* Cost curves
Equilibrium, long-run, 251,254
  monopolistic competition, 250
  monopoly, 243
  perfectly competitive markets, 68
  short-run, 68,247
Equilibrium price and quantity, 68, 240
Equity, 165,182,228,283
Escover, E.M., 188
Europe, 4,279
Existing demand, 37
Expansion path, 131
Expenses, 157,161
Explicit costs, 99
Exploitation, 58

Factor-factor relationship, 117
Factor of production, defined, 78
Factor substitution, 119
Family labor, 100
FAO, 17,18,235,275
Farm income, 151
Farm management, 141
  defined, 149
  and farm income, 151,160-163
Farm planning, 141-142
Farmer, 54,61,77,88,92,108,224, 239
Feldman, J.M., 73
Ferguson, C.E., 34,73,116
Fertilizer, 78
Flynn, J.B., 188
Fish, 77
  breeding, 3,61
  culture, 3,12
  consumption, 9,11,39,282
  production, 78,80,81,82-85,93,99
Fisheries, 1
  capture, 1
  and nutrition, 15
  trade, 16
Fishout, 272

Fingerlings
in production, 79,85,246
risk, 229
Fixed costs, 100
average, 103
short-run, 102
total, 101
Fixed input. *See* Fixed cost
Foster, T.H., 116
Fowler, E.G., 138
France, 257
Freund, J.E., 34
Fujimura, T., 96,190
Fuller, M.J., 256
Futrell, G.A., 276
Future value, of present sum, 192
series, 194-196

Gates, J.M., 116,188,190
George, T.T., 18
Gittinger, J.P., 235
Glude, B.J., 188,275
Government, intervention, 277,284
programs, 62,227,284
support policies, 287
Grass carp. *See* Carp
Greenland, D.C., 96,138
Greenland, 11
Griffin, W.L., 96,116,188,189
Gross domestic product, 16
Gross income. *See* Gross output, and Farm income
Gross margin, 153
Gross National Product, 25-26
GNP Deflator, 26
Gross output, 152,153
Grover, J.H., 18
Growth response, 84
Guevera, G., 275

Haidache, R.C., 73
Halver, J.E., 96
Hansen, G.D., 188
Hanson, J.S., 188
Hastings, W.H., 96
Hatch, L.U., 73,74,138,188,189, 275,290
Hatcheries. *See* Aquaculture; Biotechnology; and Fingerlings
Havlicek, J., 73
Havrilesky, T.M., 34
Hempel, E., 275
Herman, M., 275
Hirshleifer, J., 34
Hjul, P., 18
Hopkin, A.J., 201,234
Hopkins, K., 189
Hopkins, M., 189
Horizontal integration, 270
Hottlet, P., 275
Houston, J.E., 73
Huang, H., 189
Hutchings, D., 189

Imperfect competition, 244. *See also* Monopolistic competition
Imperfect substitution, 128
Implicit costs, 99
Implicit GNP deflator. *See* GNP deflator
Income, 38
Income distribution, 283
Income effect, defined, 41
Income elasticity of demand, defined, 49
calculation of, 50. *See also* Demand; and Elasticity
Income statement, 156
Income taxes, 154
Increasing returns to scale, 112-113. *See also* Returns to scale
Index numbers, consumer price, 27
Laspeyre index, 291,293
Paasche index, 293
India, 278
Indonesia, 197,278,285
Inferior goods, 50
Inflation, and economic risks, 222

Input substitution, marginal rate of technical substitution, 119
Inputs, average product of, 91
changes in price of, 60
combining, 117
costs, 99-104
defined, 79
indivisibility, 94
isoquant curve, 119
long-run, 109
marginal product of, 89,91
one variable, 91
policy effect, 223
two variables, 118
Insurance, 225
Integration, 270
Interest rate, 191,208,211
Internal rate of return, 208-209, 211-213,215-216
Isocost, defined, 124
calculation and slope, 125
curve, 124
Isoquant, complementary goods, 130
defined, 119
imperfect substitutes, 128,129
profit maximization with two inputs, 126
Isorevenue, 134-135
Israel, 285
Isvilanonda, S., 116,189
Italy, 280
Ivers, T.C., 73

Jamaica, 61
Japan, 3,16
Jenkins, W.E., 97,188
Jensen, J.W., 188,235
Jentoft, S., 17
Johns, M.A., 116,189
Johnson, W.E., 116,187
Jolly, C.M., 18,138,189,290

Kabir, M., 73
Kahma, I.H., 73
Karp, L., 96
Katoh, J., 189
Kensler, C., 18
Keown, A.J., 234
Kingsley, J.B., 74
Kinnucan, H., 74,138,234,256,275
Kleinfelter, D.A., 235
Kleith, W.R., 74
Klemetson, S.L., 189
Knoght, L.H., 95
Korea, 16
Krantz, G.E., 96

Labor, as factor of production, 78
return to, 164
family, 100
in natural monopoly, 246. *See also* Supply curve
Land, 78,143
Land use, 144
Lardner, R.W., 34
Lartey, B.L., 18
Laspeyre index, 291
Latin America, 8
Lawrence, A.L., 116
Law of demand, 36
Law of diminishing returns, 83
Least-cost combination of inputs, 125
Lederig, D., 189
Leeds, R., 189
Lent, 39
Leopold, M., 189
Leung, P., 190
Liabilities, 182,183. *See also* Balance sheet
Liao, D.S., 189,275
Lightner, D.V., 290
Lin, B., 74,275
Lins, D.A., 235
Lipschultz, F., 96
LiPuma, E., 18
Lizama, L.C., 138
Lizarondo, M.S., 96,188
Loans, 182

Logan, S.H., 116,190
Long, C.L., 96,138
Long-run, cost, 109-112
  defined, 109
  profit maximization, 241
Long-run supply, 62
Louisiana, 7,40
Lovell, R.T., 73,97
Lovshin, L.L., 189
Luke, D.B., 189
Luzar, E.J., 95,256

Macroeconomics, 24,25
Malaysia, 278,286
Management, 149,150
Marginal cost, curve, 102
  defined, 102
  long-run, 110-112
  profit maximization, 106. *See also* Cost curves
Marginal physical product
  calculation, 86
  curve, 86
  defined, 85
  relationships of TPP, APP, and MPP, 87
Marginal rate of product substitution, 135
Marginal rate of substitution, 122
Marginal rate of technical substitution, 119-123
  diminishing, 120-121
Marginal revenue, calculation of, 54,245
  defined, 54,241
  profit maximization, 106-107
  relation to elasticity, 55
  relation to supply curve, 241
Marginal revenue curve, for monopoly, 245-246
Marginal value product, 90-91
Mariculture, 3,280
Marion, J.E., 139
Market, 25,28,29,237
Market demand, characteristics of, 36
  defined, 35,57
  relation to marginal revenue, 245
  under monopoly, 245
  under oligopoly, 248
  under pure competition, 241. *See also* Demand
Market determined prices, 240
Market equilibrium, 68
Market functions, 258
Market information, 264
Market intelligence, 264
Market patterns, 237
Market period, 67
Market structure, 237
Marketing, 257
  channels, 264-267,272
  defined, 259
  efficiency, 258
  function, 258
  function of management, 147
  in risk management, 224
  margin, 57,267
  risks, 222
  utilities, 260
  technology, 61-62
Marr, A., 18
Martin, J.D., 235
Martin, N.R., 188
Maurice, S.E., 34,73,116
Maximization of profit. *See* Profit
McCoy, E.W., 187,276
McCracken, V.A., 74
McDonald, C.R., 116
McDowell, L.R., 139
McGuigan, J.R., 201,235
McLarney, W., 138
McVey, J.P., 116,188
Mead, C., 290
Mead, J.W., 290
Meltzoff, S.K., 18
Menasveta, P., 18
Mexico, 258,279
Microeconomics, 24

Milkfish, 12
Miller, F., 18
Miller, J.D., 256
Miller, J.S., 276
Mims, S.D., 276
Mississippi, 6,61,224
Monopolistic competition, 250-253
Monopoly, bases of, 244
  defined, 243
  demand-supply analysis, 245
  marginal revenue, 245
  market demand curve, 245
  natural, 246
  profit maximization, 245
Morita, S.K., 96
Morris, M., 17
Morocco, 7
Mortimer, M.A.E., 18
Moshen, A.A., 139
Moyer, R.C., 201,235
Mukhopadhyay, P.K., 96
Mulla, M.A., 139
Murray, W.G., 201,235
Musia, T., 95

National income, 27
Natural monopoly, 246
Natural resources, 16
Nelson, A.G., 235
Nepal, 146,278
Nerrie, B., 96
Net farm income, 162,163
Net national product, 26
Net worth, 182,183
Newkirk, G.F., 73
Niami, F., 73
Nichols, J.P., 116
Nieto, A., 73
Nigeria, 286
Nixon, C.J., 188
Nominal, 25
Nominal prices, 25
Normal goods, 50
Norway, 16
Nutrition, 13

Oligopoly, characteristics of, 248
Operating cost. *See* Cost and Cost of production
Opportunity cost, 23,99,163
Output, 79. *See also* Production
Oysters, 6,62,279

Paasche index, 293
Panama, 287
Panayotou, T., 96,116,189
Pappas, J.L., 235
Pardy, R.C., 116
Partial budgeting, 170
Perez, H., 189
Perfect competition, 237,250
  characteristics of pure and, 238
  monopoly distinguished, 248
Perfect knowledge, 37,239
Perfectly competitive market, 237-243. *See also* Competition
Perles, B.M., 34
Personal income, 27
Personnel, 146
Petty, J.W., 235
Philippines, 278
Pigott, G.M., 276
Planning horizon, 109
Poe, W. E., 139
Point elasticity, computation, 45,46. *See also* Elasticity
Policy, 277,285,287
Pollard, B.J., 116
Pomeroy, R.S., 189
Positive economics, 24
Prawns, 41
Preference, 22
Present value, 191
  calculation, 205
  net, defined, 205
  of a future payment, 196
  of series, 198-201,205-211. *See also* Discounting

Price. *See also* Demand; Supply; Elasticity; Nominal price; Relative price
Price ceilings. *See* Price controls
Price competition, 248
Price controls, 277,285
Price cross-elasticity of demand, 51
  coefficient, 51
Price determination, 268
Price elasticity of demand,
  coefficient of, 43,44
  determinants, 48. *See also* Elasticity
Price flexibilities, 67
Price index, 292
Price leadership, 249
Price supports, 285,287
Price takers, 92,238,242
Prochaska, F.J., 74,275
Product transformation curve, 134
Products, complementary, 136
  competitive, 136
  joint, 136
  supplementary, 136
Product-product relationship, 117, 134
Producer supply, 58
Production, aquacultural, 284
  cost of, 100-105
  defined, 77
  economic region of, 1,5,132
  elasticity, 94,281
  examples of, 1
  factors of, 78
  fish, 78
  limitations, 287
  long-run organization of, 109-112
  optimal combination of resources, 90
  rational area, 133
  risk, 223
  three stages of, 88
  transformation, 134
  two outputs, 133
Production function, 79,92,102
  average physical product, 86
  average physical product curves, 86,88
  defined, 79
  marginal physical product, 86
  marginal physical product curves, 86,88
  total physical product, 86,87,88
Production-possibility curve, 134
Profit, 91,92,106,154
Profitability, 163,212
Profit and loss statement, 156
Profit maximization, 125,129,241
Pullin, R.S., 138
Pure competition, 237,240,250

Quantity demanded, 37
Quantity supplied, 59,63

Rabanal, H.R., 190
Rajendren, R.B., 188
Rameses, S., Jr., 276
Ratios, 184-187
Rauch, H.E., 96,190
Raulercon, R., 256
Real estate, 149
Real income, 293
Relative price, 37
Resource inventory, 142
Resources, 77,239,244
  allocation of, 61
  optimal combination of, 90. *See* Allocation of resources
Returns to equity, 165
Returns to labor and management, 164
Returns to scale, 113
Revenue function. *See* Marginal revenue
Reynolds, L.G., 34
Rhodes, R.J., 190
Richardson, J.W., 188
Ridler, N.B., 73

Risk, types, 221-223
  management, 221,224
  measures, 228
  utility, 232
Risk aversion, 285,287
Roberts, K.J., 190
Robinson, K.L., 74
Rogers, G.L., 189
Rouse, D.B., 139,189
Rupp, E., 18
Rwanda, 238
Ryther, J.H., 138

Sadeh, A., 96
Salmon, 16,118,257
Salon, O.T., 188
Salvage value, 166,167,169
Samples, K.C., 190
Sandifer, P.A., 18,97,187
Scarcity, 21,24
Schkade, L.L., 34
Schroeder, G., 139
Schupp, A.R., 73,256
Schuur, A.M., 116
Schwartz, N.B., 189
Schwedler, T., 189
Scott, D.F., Jr., 235
Seng Keh, T., 97
Shang, Y.C., 18,74,96,190,235
Shaw, S.A., 276
Shehadeh, A.H., 139
Shellfish, 3,16,40,58
Shepherd, G.S., 276
Shigekawa, K.J., 116,190
Shleser, R.A., 96,190
Short-run, 100-104,108. *See* Cost; and Production
Short-run equilibrium. *See* Equilibrium
Short-run supply curve, 67,240
Short-run supply in monopoly, 247
Shrimp, 3,9,16,47,50,51,143,257, 279
Sierra Leone, 278
Simon, J.L., 34,256
Sinclair, H.M., 17
Sindelar, S., 74,189,234,275,290
Smith, I.R., 97,139,188,276
Smith, T.I.J., 97,187,189
Social cost of production, 99
Spivey, W.A., 97
Spreen, T.H., 275
Stamp, N.H.E., 190
Starr, P.D., 18
Stauffer, J.R., Jr., 17
Stickney, R.R., 18,95,97
Street, D.R., 18,74
Stokes, A.D., 97
Stokes, R.L., 190
Substitute goods, availability, 48
  elasticity factor, 48
Substitution, imperfect, 128
  constant, 129
  product-product, 134. *See* Elasticity of substitution
Substitution effect, 48,51
Sullivan, G.M., 74,256,276
Superior goods, 50
  defined, 50
  substitution effect to, 51
Supply, 21,58
  market, 17
  short-run, 67
  producer, 58
  long-run, 112. *See also* Demand and supply analysis
Supply conditions, 58,59
Supply curves, 59-61,66,67,241. *See also* Cost curves
Supply of fish, 58,60
Swinton, S., 275

Talpaz, H., 97
Technology, 58,60,109,111
The Cameroon, 287
The New Encyclopedia Britannica, 18
Thia-Eng, C., 97
Third World, 1,10,281
Thorp, T., 275

Time preference, 192
Time value, 191
Tilapia, 7,50,59-62,118,145,146, 171,213
Tokrisna, R., 96,116,189
Topography, 143
Total cost, 90,99-110,154,175,241
defined, 90,99-100
Total physical product, 86-88
Total revenue, 54,55,70,90,107, 241,246. *See* Average revenue; and Marginal revenue
Total revenue curve, 56
Total variable cost, 101-102
Tomek, G.W., 74
Toussaint, W.D., 138
Trotter, W., 256
Trout, 16,41
Tsur, Y., 97
Tsvilanonda, S., 96
Tunisia, 7

Unitary elasticity of demand, 55,56
United States, aquacultural production, 5-7,118,257,270
world trade, 16
demand for fish, 38,40,41
government programs, 227,257
USDA, 18,74
USDI, Fish and Wildlife Service, 97
Utility, 232

Value of marginal product, 90,92
Van Dam, A.A., 97
Van Der Lingen, I., 18
Van Eys, S.S., 275
Vandeveer, L.R., 116
Variable costs. *See* short-run
Variable inputs, 101-102
Varley, R.L., 190
Venezuela, 146
Vertical integration, 270
Vondurska, J., 276

Waldrop, J.E., 116,256,276
Walker, N.P., 190
Warren, L.F. 201,235
Water supply and quality, 279
Wattanutchariya, S., 96,116,189
Warm water, 5
Weston, F.J., 201,235
Whetham, E.H., 276
Williams, H.R., 34
Williams, N.A., 74
Williams, S.B., 188
Willis, S.A., 97
Wilson, R.P., 139
Wineholt, D., 74,275,290
Wholesale price index, 28
White Amur. *See* Carp
Wonnacott, P., 34

Young, B.T., 73

Zidak, W., 74

Printed in the United States
by Baker & Taylor Publisher Services